2

A Little
Book *of*
Latin *for*
Gardeners

Also by this author

Housman Country: Into the Heart of England

A LITTLE
BOOK *of*
LATIN *for*
GARDENERS

Peter Parker

Little, Brown

LITTLE, BROWN

First published in Great Britain in 2018 by Little, Brown

1 3 5 7 9 10 8 6 4 2

A CIP catalogue record for this book
is available from the British Library.

ISBN 978-1-4087-0616-9

Typeset in Arno by M Rules
Printed and bound in Great Britain by
Clays Ltd, Elcograf S.p.A.

Papers used by Little, Brown are from well-managed forests
and other responsible sources.

MIX
Paper from
responsible sources
FSC® C104740

Little, Brown
An imprint of
Little, Brown Book Group
Carmelite House
50 Victoria Embankment
London EC4Y 0DZ

An Hachette UK Company
www.hachette.co.uk

www.littlebrown.co.uk

for
Thomas

CONTENTS

HISTORY

Man, says Montaigne, is undulating
and diverse. He didn't know the
ways of horticultural nomenclature,
lucky Montaigne – for anything more
undulating and diverse than these no mind
could conceive.

Reginald Farrer, *My Rock Garden* (1908)

'Oh, the agony of not knowing Latin!' a character in one of Barbara Pym's novels exclaims as she attempts to decipher the inscription on a church monument. Few people these days would regard not knowing Latin as any sort of deprivation. Most of those who were obliged to learn Latin at school tended to regard it as a form of endurance rather than something that would usefully broaden their minds. Generations of schoolchildren have gloomily chanted to themselves the anonymous lines

Latin is a language
As dead as dead can be;
First it killed the Romans,
And now it's killing me.

Latin does, however, still live and have one very particular function used by millions of people with no classical education at all. It remains the universal language – the *lingua franca* – of the international gardening community. Botanical Latin became the language in which plants were named and identified because the categorising of plants began in the ancient world, and because Latin was the common language by which scholars communicated with each other during the period natural history took off as a new science. Although it may be useful to have learned some Latin at school, botanical Latin is not the same thing as the language in which you may have read about Caesar's Gallic Wars. The kind of Latin used in the naming of plants does indeed date back to the Romans, and to Pliny the Elder in particular, whose encyclopaedic *Historia Naturalis*, written between A D 77 and 79, remained a principal source for those writing about plants well into the sixteenth century. Botanical Latin is not, however, the same language as that used in ancient Rome. Indeed, it is less a language than an international tool, largely invented in the renaissance and refined and more or less standardised in the eighteenth century by the Swedish botanist Carl Linnaeus. Many of its words are derived from Greek, while others refer to things – geographical and scientific discoveries, for example – that would not have been known in the ancient world. The good news is that unless you are a professional botanist you don't really need to know much about Latin grammar, nor should you worry too much about pronunciation. It's as well to learn that a cotoneaster isn't pronounced as if it were two English words referring to thread and a Christian festival ('cotton-Easter'), but many of the finer points of botanical Latin pronunciation are disputed. When I first became interested in gardening I was at a plant sale where I enquired about an eryngium, which I pronounced with a soft 'g'. A bossy upper-class woman of the sort often found at such sales 'corrected' me by repeating the word with a hard 'g'. If this kind of thing worries you, then you can read the section on pronunciation in William T. Stearn's magisterial *Botanical Latin: History, Grammar, Syntax, Terminology and Vocabulary* (1966, revised 1983); but as Stearn himself concedes: 'How [names] are pronounced really matters little provided they sound pleasant and are understood by all.' And the notion of a language understood by all is the

Sea holly, *Eryngium* – hard 'g' or soft 'g'?

very foundation of – and is the reason for – botanical Latin.

It is, of course, quite possible to cultivate a garden without knowing the Latin names of any of the plants in it. However, you will find that most nurseries give plants their proper botanical names when listing, labelling, displaying and selling them. At the most basic level, therefore, it is useful to know that if you are looking for a hollyhock you will need to start your search not under 'H' but under 'A' for *Alcea*. One of the principal reasons for calling a plant an aquilegia rather than a columbine or granny's bonnet, or any of the many regional names, is that the botanical name can be understood wherever you happen to be – in any county in England and in any country in the world. Rather than mystifying each other with their holly, their *houx*, their *Stechpalme* and their *agrifoglio*, British,

French, German and Italian gardeners can all use the Latin word *ilex* and all know what they mean. This also sorts out the muddles that arise when people use the same English name for totally different plants: such as 'geranium' to describe not only hardy members of the *Geranium* genus (sometimes known as cranesbill), but also tender members of the *Pelargonium* genus, in particular the scarlet 'geraniums' that adorn countless windowsills and public bedding-schemes. There is the additional pleasure for the curious-minded in learning that 'cranesbill' is derived from the botanical name: *geranos* is the Greek word for 'crane', whose long beak the plant's elongated mature seedheads supposedly resemble. (Take a look at the woodcut illustration of *Geranium moschatum*, 'Musked Cranesbill', in Gerard's *Herball* and you will see a nestful of cranes with lots of upward-pointing 'beaks'.) There is, however, further confusion for gardeners: while *pelargos* is the Greek for 'stork' – pelargoniums too having elongated seedheads – the English name storksbill is more often used for erodiums, whose name derives from *erodios*, the Greek word for 'heron'. Americans are in this instance more ornithologically accurate than the British, and in the United States erodiums are sometimes called heron's bill, although *Erodium gruinum* (from the Latin *grus*, a crane)

Gerard's 'Musked Cranesbill',
Geranium moschatum

adds further ornithological muddle and is known as the long-beaked stork's bill. I'm not sure that I could, without carefully studying images, detect the differences between the beaks of these birds (which, incidentally, belong to three different zoological families), but the pioneering English botanist William Turner (?1508–1568) probably could have done, since he also wrote a treatise on birds – in Latin. It may be some comfort to know that, unlike the birds, all three plants are members of the same botanical family, the *Geraniaceae*.

If all this seems bewildering, imagine a nursery where plants are arranged any old how rather than grouped alphabetically by genus and species. That is how the horticultural world once looked until such men as Theophrastus, Andrea Cesalpino, William Turner, Conrad Gessner, Matthias de l'Obel, John Ray and Carl Linnaeus attempted to introduce order into the plant kingdom. Linnaeus apart, these names will be unfamiliar to most gardeners, but they belong to people who devoted long years to the study of plants in an attempt to work out how they were related and how they might be differentiated. Largely thanks to them, plants are now taxonomically identified by several strata: Division (now sometimes called Phylum), Class, Order, Family, Genus, Species. The gardener and general reader can leave the first three of these to the botanists. It is not even very important to know about plant Families, which like human ones can be large and diverse (the *Ranunculaceae* include plants as apparently different as aquilegias, clematis, delphiniums, hellebores and buttercups) or small and exclusive (the *Eucryphiaceae* consist solely of eucryphias). If you visit a botanical garden you may well find that the beds are arranged by Family, which allows you to marvel at the way in which unexpected plants are related to each other, but nurseries and their catalogues customarily arrange plants simply by genus and species, a two-word designation regularised by Linnaeus and called a binomial.

The genus (plural, genera) is rather like a surname, and is followed by a further defining name equivalent to a forename and known as the 'specific epithet' because it tells you the species. For example, there are many species of *Achillea*, of which *Achillea millefolium* is the one most often grown by gardeners. There can be further subdivisions within species, such as varieties, subspecies and cultivars, and these are described and explained in 'Essentials'.

The binomial system is not too hard to grasp, but plant names have a nasty habit of changing. Scientific advances such as the discovery of DNA have made it possible to recognise that plants previously thought to be of the same genus are not in fact related and so have to be renamed. While much of this is really of interest only to botanists, it has a considerable impact on gardeners who want to label their plants correctly, or even find them in a nursery. To take a quite recent example, and a plant most gardeners know well: all Michaelmas daisies used to be members of the *Aster* genus, but in 2015 were widely redistributed amongst other genera, with only the southern European *Aster amellus* (Italian aster) and its close relatives retaining the name. Many of the other species, including the popular *A. laevis* and *A. novi-belgii*, were renamed *Symphyotrichum*, while several more, such as the pretty, dark-stemmed, low-growing *A. divaricatus*, became a *Eurybia*. (This last in fact became *E. divaricata*, the change of genus name requiring a change of gender for the specific epithet – see 'Essentials'.) Others became *Galatella, Tripolium, Doellingeria, Oreostemma, Xanthisma* and *Ampelaster*. All that said, several years after these changes you will still find these plants all listed under *Aster* in nurseries, which naturally enough prioritise the convenience of their customers over advances in science. Some nurseries do encourage us to catch up. In my own garden recently I found a nursery label which read: '*Ageratina altissima* "Chocolate" (was *Eupatorium rugosum* "C")'. I have to confess that I still think of the plant as a eupatorium, and am not one of those people who make new labels when something I have been growing for several years is suddenly reclassified. Similarly – and perhaps less excusably, since these changes are much older – I still refer to my beautifully arching *Rosa glauca* as *Rosa rubrifolia* and sometimes call the invasive but irresistible plume poppy I was given by an elderly friend long since dead *Bocconia* (which was how she knew the plant) rather than *Macleaya*, as it is now correctly named.

Because so many English words in everyday use are derived from the Latin, some botanical names hardly need decoding, and specific epithets such as *odoratus* or *grandiflorus* are unlikely to tax anyone too severely. Other names, however, may seem obscure, confusing or occasionally inexplicable even when you know the derivation – but that to me is part of the fascination. One of the aims of this book is to show that botanical Latin should be regarded as entertaining rather than off-putting, and to this end there will be plenty of stories about how plants got their names. So, for example, *Eupatorium* was named after Mithridates Eupator, King of Pontus and Armenia Minor between 120 and 63 BC, who will be described in the chapter on eponyms – as will Paolo Boccone and Alexander Macleay, who gave their names successively to the plume poppy. For practical horticultural purposes you don't, of course, need to know any of this – or know about plants named after animals, birds, insects and parts of the body – but for anyone of a curious turn of mind, it adds to the pleasure of looking at plants to know a little about how and why they came to be named. When I see the pale blue flower-buds of the striped squill emerge from the February soil, I recall that the plant is called *Puschkinia scilloides* after the aristocratic and gloriously named Russian mining expert Count Apollo Mussin-Puschkin (1760–1805). The Count led a botanical expedition to the Caucasus in 1800 and sent this bulb, along with specimens of such other plants as the gracefully arching *Campanula alliariifolia* and the white-flowered *Rhododendron caucasicum*, to Sir Joseph Banks at Kew Gardens, whom he had met while serving as Russian Minister in England. It doesn't make the plant any prettier, but it does conjure up the great era of plant hunting, when thousands of plants we take for granted in our gardens were introduced to Britain from distant corners of the globe.

Of more practical use are specific epithets that refer to the physical characteristics of the plant, such as *altissimus* (growing very tall) and *rugosus* (wrinkled) – indicating respectively habit and leaf type – or names reflecting coloration, such as *rubrifolius* (red-leafed) and *glaucus* (greyish blue-green). Similarly, some species take their names from the other ways that we appreciate them – touch, smell and taste – useful information when you are deciding what to grow and where. Names identifying the geographical region in which a plant originated can also give valuable clues to how it might or might not flourish in your own garden.

But first some history.

The naming of things is a fundamental human occupation, helping us to make sense of the world that surrounds us. From the moment of creation, according to the Book of Genesis, God brought every beast of the field and fowl of the air to Adam 'to see what he would call them'. Once named, things need to be placed in categories, and around 300 B C the Greek philosopher Theophrastus embarked on the job of naming and sorting some 500 plants. Theophrastus was not only a philosopher; he was also a gardener, having inherited the botanical garden that his teacher Aristotle had created at Athens. This meant that he had plenty of plants to scrutinise while taking his daily constitutional, and he first of all divided them into four broad categories: trees, shrubs, subshrubs and herbs. He then studied the individual plants, taking particular note of such attributes as the shape of their leaves, the kind of root, fruit or bark they had, and how and where they grew, noting both their similarities and differences. It may seem odd to us today that he paid little attention to flowers, but these were regarded as a type of leaf for another couple of millennia: it wasn't until 1690, when the English naturalist John Ray published a new method for the identification of plants, that the word petal came into existence. Theophrastus's pioneering writings on identifying plants more or less disappeared until they were discovered by chance and translated into Latin in the fifteenth century. They were among a large collection of classical texts held in the Vatican Library, and in 1451 Teodoro of Gaza, a Greek scholar based in Italy after whom the gazania is named, was summoned to Rome by Pope Nicholas V to translate into Latin both Theophrastus's work and Aristotle's similarly innovative treatise on animals. These were published respectively as *Historia plantarum* (1483) and *Historia animalium* (1476).

So the first work identifying and naming plants was in fact written in ancient Greek, which explains why we have pelargoniums and geraniums, and indeed anemones, peonies and antirrhinums – all names derived from the Greek. And while we are about it, it is worth mentioning that Paradise, the word used for the very first garden, is derived from the Greek word *paradeisos* meaning a park or orchard, which is itself derived from words in the Indo-Iranian language of Avestan: *pairi* = around + *daēza* = wall.

Because Theophrastus's work had for so long disappeared from view, the Roman writer Pliny the Elder was widely regarded as the earliest author to attempt to make sense of the natural world, in his ten-volume *Historia Naturalis*, which he left incomplete when he perished in the AD 79 eruption of Vesuvius. Pliny had gone to investigate a cloud emitting from the volcano that was, perhaps appropriately, shaped like an umbrella pine (*Sciadopitys verticillata*). Botanising indeed often takes people to remote and dangerous places, and many later naturalists and plant collectors would die in pursuit of their calling.

Classicists will recognise that the binomial for the umbrella pine is in fact mongrel: the genus takes its name from the Greek *skiadeion* (parasol) + *pitys* (fir), while the species name is taken from the Latin *vertex* (gen. *verticis*) (whirlpool), referring to the way the leaves form a ring around an axis. (See 'Essentials' for an explanation of the two forms of the noun.) Pliny borrowed heavily from his Greek predecessor; like many of the early writers on plants, he was more of a compiler of information than a generator of it. Where Latin names existed, he used them to replace Theophrastus's Greek ones, but when they did not he merely transliterated the Greek names and altered their endings to bring them into line with his own language, so that, for example, *narkissos* became *narcissus* and *krokos* became *crocus*. With other word-endings there was less consistency: the Greek -*on* tended to be transliterated as the Latin -*um*, but sometimes remained unaltered, as in *rhododendron*. Gardeners, unlike botanists, need not concern themselves with all this, but

'The white dwarfe Cistus of Germanie', a rock-rose of the Cistus genus

it does explain why botanical Latin often looks different from classical Latin. It should also be stated from the outset that a considerable number of genus names have no 'meaning', but are simply what the plants were called by classical authors, handed down across millennia. This is particularly true of trees such as oaks (*Quercus*), walnuts (*Juglans*), planes (*Platanus*), elms (*Ulmus*), willow (*Salix*) and yew (*Taxus*), but also of many other familiar plant names, such as *Cistus*, *Clematis*, *Geum*, *Helleborus*, *Lathyrus*, *Papaver*, *Rosa* and *Thymus*.

More Greek names were written down by Dioscorides (c. AD 40–90), a Greek physician who served with the Roman army and whose five-volume encyclopaedia of herbal medicine became another major source for later botanists, circulating in its original Greek and translated into both Latin (as *De Materia Medica*) and Arabic during the medieval period. It is one of the principal disadvantages of early herbals that plants were identified for their medicinal value, which meant that anything that could not be used to treat or cure illnesses tended to be sidelined. Of course it was important that medicinal herbs should be identified correctly by physicians. These herbs were often supplied by women who gathered them in the countryside and sold them in cities. Although many of these 'simple gatherers' knew their herbs and were wholly trustworthy, a physician unable to recognise one plant from another might be palmed off with worthless weeds, or – even worse – with misidentified toxic ones.

It was not until 1544 that the Sienese botanist and author Pietro Andrea Mattioli (1501–77), in his *Discorsi* or commentary on Dioscorides, wrote about plants with no medicinal properties. Like many writers of the time, Mattioli Latinised his surname and is often referred to as Matthiolus; the *Matthiola* genus of plants that includes Brompton and night-scented stocks was named in his honour.

Mattioli recorded 100 new plants in his *Discorsi*, but, just as many literary academics today spend more time writing about what other scholars have written about a text than about the texts themselves, so early botanists were often content to follow the authorities from the ancient world rather than look afresh at the plants that had been described. It was much the same with illustrations: those who copied the text of manuscripts also copied the drawings, however inaccurate they may have been. It is clear from medieval Books of Hours and other manuscripts that people were capable of producing accurate representations of plants, though these were often stylised for decorative effect or painted for symbolic rather than botanical purposes. Among manuscripts on botanical topics, the Carrara Herbal, created in Padua between 1390 and 1400, and the Codex Bellunensis, created at Belluno in the Veneto between 1400 and 1425, seem to show plants painted from life rather than merely copied from earlier illustrations. In the latter work, some plants are shown as they would be in modern botanical drawings, with supplementary illustrations of roots or bulbs and seed heads, all of which would aid identification.

Books on plants began to appear soon after the invention of the movable-type printing press in the mid-fifteenth century. The thirteenth-century *Liber de proprietatibus rerum* by the French scholar Bartholomaeus Anglicus, published in a printed edition around 1470, was an attempt to identify 'The Properties of Things' and included in its nineteen books one titled *De herbis et arboribus* in which plants and trees and their medicinal properties are listed in alphabetical order. The book proved very popular and went through twenty-five editions by the end of the century. Published in Augsburg some five years later, Konrad von Megenberg's *Das Buch der Natur* devoted one of its eight chapters to plants and included the first known botanical woodcuts. These two woodcuts are fairly crude, but it is perfectly possible to recognise a lily, a violet, a pink, a lily of the valley and some kind of gourd among the plants depicted.

In England in 1538 William Turner published *Libellus de Re Herbaria*, in which he provided 'The names of herbes in Greke, Latin, English, Duch [i.e. German] and Frenche wyth the commune names that Herbaries and Apotecaries vse'. These were, however. still arranged alphabetically rather than in groups. Turner's later designation as 'the father of English botany' is based instead upon two later works: *The Names of Herbes*, published in 1548 as the first original herbal in the English language and listing many British plants for the first time, and *A new herball* (1551–64), also written in English but with the plants arranged in alphabetical order by their Latin rather than their English names, from *Absinthum* ('Wormwoode') to *Vicia faba* ('the Beane'). Turner had studied plants in the field, not only in Britain but also in Europe, where he was frequently exiled owing to his staunch and vigorously expressed Protestant views. (Alongside his works on natural history, he published such books as *The huntying and fynding out of the Romyshe foxe* and *The Huntyng of the Romyshe Vuolfe*, the latter including illustrations in which wolf-headed bishops are depicted assisting at the burning of Latimer, Ridley and Cranmer.) He also wrote (in Latin in 1544) the first work of ornithology in England, which included descriptions of birds given by Aristotle and Pliny supplemented with his own, often very vivid, observations of those he had studied in the wild.

A very good reason for the use of Latin in scholarly books was that it was a stable language at a time when English, for example, was not. When Turner was writing his books in the 1530s and 1540s, the language had moved on considerably from the Middle English of Chaucer's time, 150 years earlier. As for the spelling of written English, it was not standardised until centuries later, so the modern reader stumbles over such sentences as 'The Poticaries of Colon before I gaue them warning vsed for this the bowes of vghe.' It takes a while to grow accustomed to 'floures' and 'leves' and the occasional 'litle schrube', and to place-names that seem at first unrecognisable and often lack an upper-case first letter – 'bon and colon' (the latter being Cologne rather than the city in Panama we actually now spell like that).

Identifying plants accurately was one thing, but ordering them was quite another. Various attempts were made, notably by the German doctor and Lutheran pastor Hieronymus Bock (1498–1554), whose *Neue Kreütter Buch*, or 'New Plant Book', published in 1539, replaced the alphabetical listing used in most herbals with his own groupings. 'I have placed together, yet kept distinct, all plants which are related or connected, or otherwise resemble one another and are compared,' he wrote.

The Swiss naturalist Conrad Gessner (1516–65), whose four-volume *Historia animalium* laid the foundations of modern zoology, worked on a companion project, *Historia plantarum*, which arranged plants into groupings according to their flowers, seeds and fruit, and also used the terms 'genus' and 'species'. However, it remained unpublished for several centuries after his early death from plague. The first person to publish a properly systematic grouping of plants, therefore, was the Italian botanist Andrea Cesalpino (1519–1603), whose *De* plantis *libri XVI* (1583) arranged 1,500 plants into 32 groupings, also largely according to their fruit and seeds. Cesalpino was director successively of the botanical gardens at Pisa (founded 1547) and Rome (founded 1566), and in 1563 was one of the first people to create a herbarium or collection of dried plant specimens. The herbarium was sometimes called a *hortus siccus* (literally 'dry garden'), but unlike a real garden, or those pretty albums of pressed flowers created by the Victorians, botanical specimens in a herbarium were arranged for scientific use rather than to give pleasure, often on cards for easy storage and reference. In Cesalpino's herbarium the specimens were laid out in groups, with Greek, Latin and Italian alphabetical indexes, and what is more Cesalpino differentiated between species in the same genus, defining them by Latin adjectives describing appearance or habitat and so looking forward to Linnaeus's binomial system. Herbariums would become invaluable databases, allowing botanists to compare and classify newly discovered plants. For example, in 1771 Joseph Banks arrived back in England from his expedition on the *Endeavour*, bringing with him 30,000 dried specimens representing more than 3,600 species, almost half of which were new to Europe.

In 1597 one of the best loved botanical books in the English language was published: John Gerard's *The Herball*

John Gerard, author of *The Herball*

or *Generall Historie of Plantes*. Gerard (c. 1545–1612) had been apprenticed to a barber-surgeon and been admitted to their Company in 1569, but he was chiefly known as a herbalist. Although he has been variously accused of being a crook, plagiarist, blunderer, dupe and liar, Gerard knew a great deal about plants because he actually grew them. When the College of Physicians established a physic garden in the City of London in 1586, Gerard was appointed curator. He also superintended Lord Burghley's gardens in both London and Hertfordshire and created his own garden in Holborn. His 1596 list of the plants he grew at home was the first ever such catalogue to be published. The *Herball*, however, was largely taken from an English translation of the Flemish physician Rembert Dodoens's 1554 *Cruydeboeck*, a work that had already been widely translated and drawn upon: an English translation of a French version had appeared in 1578 as *A niewe herbal, or historie of plantes*. Gerard somehow acquired the manuscript of another translation by a certain Dr Robert Priest, which he adapted, augmented, rearranged and published as his own work. The first edition of the *Herball* was a mess, full of mistakes and with numerous illustrations appearing alongside the wrong plants. It was, however, organised according to the principles of Gerard's erstwhile friend Matthias de l'Obel (also known as Lobelius, 1538–1616), a Flemish doctor who spent much of his time in England, where he became doctor and royal botanist to James I. While working as a physician in Bristol, de l'Obel had gone with his friend Pierre Pena in search of plants in order to write their new herbal, *Stirpum adversaria nova* (1571), in which an attempt was made to classify plants by leaf type. Gerard follows this rather than Dodoens's original alphabetical listing, but de l'Obel was not best pleased with the result and would provide numerous corrections to Gerard's entertaining but distinctly slapdash book.

Indeed, the considerable reputation of the *Herball* rests on the 1633 edition, substantially revised, corrected and enlarged by Thomas Johnson, a young apothecary and plant collector who would be killed in his early forties fighting for the Royalist cause in the English Civil War. Johnson had already published two accounts (in Latin) of field trips he had made with some botanising friends to Kent and Hampstead, in which he listed the plants they had found: *Iter plantarum* (1629) and *Descriptio itineris plantarum* (1632). Two later volumes, *Mercurius Botanicus* (1634 and 1641), recorded expeditions Johnson undertook to the West Country, Wales and the Isle of Wight. These publications were the first steps towards a British Flora, a book listing and describing the wild plants growing in a particular country or district. This would have been the first book of its kind and might have been completed had Johnson survived the wounds he received when shot in the shoulder at Basing House in Hampshire. His books are not only important for recording numerous species of British wild flower for the first time, but they provide a lively account of what botanising in the field was like in the early decades of the seventeenth century. In his 'Very much Enlarged and Emended' edition of the *Herball*, Johnson tactfully writes that Gerard, in his public-spirited attempt to propagate botanical knowledge, 'endeauoured to performe therein more than he could well accomplish'. He mentions several serious blunders, but generously adds, 'Let none blame him for these defects, seeing he was neither wanting in pains nor good will, to performe what he intended'. Many people have pilloried Gerard for claiming to have seen geese hatching out of a 'barnacle tree', a long-discredited myth. It may, however, be significant that this is the final entry in his vast book, and it is possible that Gerard included it as a deliberate joke. Much less excusable is his claim that he had seen a species of peony growing 'wilde upon a cony berry' (or rabbit warren) near Gravesend. Johnson felt bound to add a note in his revised edition of the *Herball* to say: 'I have been told our Author himself planted that Peionie there, and afterwards seemed to finde it there by accident: and I do believe it was so, because none before or since have ever seen or heard of it growing wilde in any part of this Kingdome.'

'The breed of Barnacles': Gerard's little joke?

For all the book's faults, Gerard's love of flowers is apparent throughout the *Herball* in his vivid observations and descriptions. He did much to popularise an interest in plants rather than leaving it as the preserve of doctors and scientists, and while other early botanists and authors may have been more scholarly, Gerard would have been the livelier companion to take with you when visiting a garden, and he will be referred to and quoted throughout this book.

Rivalling Gerard's *Herball* in the seventeenth century were two books by John Parkinson (1566/7–1650), who ran a successful apothecary's shop in Ludgate Hill in London. Parkinson too was a well-practised gardener, as becomes clear from his *Paradisi in Sole Paradisus Terrestris, or A Garden of All Sorts of Pleasant Flowers* (1629). As well as listing nearly 1,000 plants that could be grown in the pleasure garden, kitchen garden or orchard, Parkinson provided notes on cultivation derived from his own experiences in his two-acre demesne in Long Acre, Covent Garden. (The book's title is a pun and could be translated as 'The park on earth of the park-in-sun'.) Parkinson was appointed royal botanist to Charles I, and his other major work was the *Theatrum Botanicum* of 1640. Although the plants here were divided into seventeen 'Tribes', the arrangement was that of a herbalist rather than that of a botanist, with plants grouped according to their 'vertues': 'Purging Plants', 'Breakstone Plants', 'Vulnerary or Wounde Herbes', and so on. Parkinson occasionally gives the impression of throwing in the towel – 'Strange and Outlandish Plants', 'The Unordered Tribe' – but in amongst all this are families of plants we would recognise today: 'Umbelliferae', 'Cardui & Spinosae Plantae' (thistles), 'Legumina' (pulses), and 'Gramina, Iunci & Arundines' (grasses, rushes and reeds). Within these tribes are 'chapters' grouping plants into something resembling genera, and, like Gerard's *Herball*, Parkinson's two books provide a fascinating descriptive checklist of plants whose names might have changed but which are still grown in gardens today.

There remained the problem that botanists defined plants by describing them, which meant that many Latin names were less than succinct. Parkinson lists *Plantago angustifolia sive quinque nervia major & ferrato*, for example, and as late as 1730 the East London nurseryman John Cowell in his *The Curious and Profitable Gardener* lumbered the torch thistle with the Latin name *Cereus erectus maximus Americanus Hexangularis, Flore albo radiato*. As a description this may be useful, but as a name it is absurdly cumbersome. Some attempt to reduce these lists of descriptive words to something more manageable had been made in 1623 by the Swiss botanist Gaspard Bauhin in his *Pinax theatri botanici*. Botanists might all have been using the same language, but they were also inclined to create their own names for plants, so there was no universal standard. Bauhin's work provided a concordance of 6,000 of these various names and attempted to introduce some sense of order into the prevailing chaos by recommending just two defining names of genus and species.

It was left to the English naturalist John Ray (1627–1705) to suggest a new system for naming and ordering plants. Unlike almost everyone else involved in this search for order, Ray was trained in neither botany nor medicine. Instead he came from that long and distinguished tradition of English amateur naturalists, most famously Gilbert White, who were motivated by simple curiosity about the world in which they lived. The son of an Essex blacksmith, Ray had attended his local grammar school and then gone up to Cambridge, where he eventually became a Fellow of Trinity College. In the touching preface to his *Catalogus Plantarum circa Cantabrigiam nascentium*, published in 1660 as the first ever British county Flora, he explained that he had been drawn to wild flowers while walking and riding through the local countryside in order to recuperate from a serious illness. 'I had leisure in the course of my journeys to contemplate the varied beauty of plants and the cunning craftsmanship of nature that was constantly before my eyes, and had so often been trodden thoughtlessly underfoot,' he wrote. 'Once I had become more aware of these wonders, I ceased to pass them by and treat them as matters unworthy of my attention. First I was fascinated and absorbed by the rich spectacle of the meadows in springtime; then I was filled with wonder and delight by the marvellous shape, colour and structure of the individual plants.' He decided to learn botany, but was surprised to find that there was no one at Cambridge sufficiently acquainted with the subject to teach him. Instead he taught himself, reading widely and studying plants in the field. He created a small garden at Trinity in which he could grow the specimens he had collected or been sent by others, and could note down their similarities and differences. In the *Catalogus* Ray, like Theophrastus, divided plants into the four basic groupings of trees, shrubs, subshrubs and herbs, but also allocated them to subdivisions within those categories, largely basing his ideas on those laid out by Jean Bauhin, brother of Gaspard, in his 1650 *Historia plantarum universalis*. Trees, for example, were allocated to the *Pomiferae* (such as apples and pears, citrus fruit, figs, pomegranates and others bearing stoneless fruit); *Pruniferae* (bearing stoned fruit and including olives and date-palms alongside plums, cherries, apricots and peaches); *Nuciferae* (bearing nuts); *Bacciferae* (bearing berries, including laurels, mulberry, box, myrtle, elder 'and many others'); *Glandiferae* (with acorn-like fruit, as oak, beech, cork); *Coniferae* or conifers; and *Siliquosae* (bearing

pods: 'Laburnum, Judas tree, Pudding-pipe tree'). This left
out a great many other trees – 'some bear only catkins like the
Birch and Willow, etc., while others have a seed enclosed in a
membrane like the Ash, Maple, Elm, Lime, etc.' – but it was a
start. Shrubs he divided into two groupings: '*Spinosi* such as
Barberry, Christ's Thorn, Buckthorn, Gooseberry' and '*Non
Spinosi* such as Broom, Alder, Jasmine, Privet, Chaste tree'.
He added that shrubs 'can also be divided into flower-bearing,
fruit-bearing and climbing', and that subshrubs were 'a simple
and not very numerous family' of mostly scented plants
'such as Hyssop, Lavender, Savory, Sage, Poley [teucrium],
Southernwood, French Lavender'. Ray found herbs 'so numerous that it is difficult, in fact almost impossible, to arrange
them into definite classes so that no plant belongs to more
than one class, or is classified ambiguously'. He nevertheless
listed twenty-one families, some of which are still recognisable
today, though their names may have changed: *Umbelliflorae*,
for example, is now the *Umbelliferae*, and his *Graminifoliae* is
the *Gramineae* or grass family. Ray ends this outline classification by stating that: 'Plants can also be classified by various
other methods e.g. by the nature of their roots, stalks, flowers,
seeds or leaves etc., but a detailed discussion of these topics
is beyond the scope of my book.' He would, however, expand
on this in his *Methodus plantarum nova* (1682), in which he
suggested classification by a close study of plant structure.
Ray would go on to compile a *Historia Plantarum*, published in
three large volumes (1686–1704) running to over 2,000 pages.
Here he listed some 6,000 plants, arranged not alphabetically,
as in the *Catalogus*, but grouped into 125 families, and this
work laid the foundations of modern botanical nomenclature.
Because Ray wrote in Latin, and his books were unillustrated,
he did not get his due in Britain, where far less reliable volumes

such as Gerard's *Herball,* written in English and plentifully illustrated with woodcuts, had a much wider circulation.

The Swedish botanist and explorer Carl Linnaeus (1707–78) also published in Latin rather than his native language, but his well-honed propensity for self-promotion, meant that in Europe his books swiftly eclipsed the pioneering work of Ray. Linnaeus was by all accounts a dreadful man – vain, boastful, untruthful – and his notion that plants could be 'sexually' differentiated simply by studying the number and arrangement of their reproductive parts proved a dead end. We must nevertheless be grateful to him because in his ever expanding and much revised and corrected *Systema Naturae* (notably in the tenth edition of 1758), he regularised and popularised the binomial system of genus and species we use today to identify all living things. And it was with plants that Linnaeus developed the

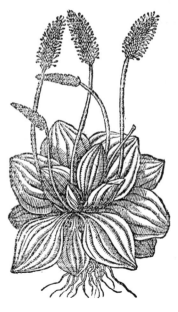

binomial system in his *Species Plantarum* (1753). This book, running to 1,200 pages, listed almost 5,900 species of plant (of the mere 10,000 Linnaeus incorrectly imagined to exist), arranged into 1,098 genera but with no descriptions. Linnaeus recognised the need among botanists for names that described a species in some detail, but the lengths such names reached would hardly do as handy labels, for which there was a still greater need. He therefore devised a specific epithet for everyday use, and

Hoary plantain, *Plantago media*

this is what has survived, while his 'diagnostic' phrases would gradually disappear. Linnaeus's binomial system has proved invaluable and enduring, neatly reducing the name of the hoary plantain, for instance, from the interminably descriptive *Plantago foliis ovato-lanceolatis pubescentibus, spica cylindricali, scapo tereti* (plantain with downy narrow oval leaves tapering to a point at each end, a cylindrical head and a smooth stem) to the succinctly defining *Plantago media* (intermediate plantain).

There are in fact some 200 species of the humble plantain, the genus name of which comes from the Latin *planta*, meaning the sole of the foot, + -*ago*, a suffix meaning 'like', because the leaves of some species tend to lie flat against the ground. The ones most likely to appeal to the gardener are cultivars of the broad-leaf plantain, *Plantago major*, notably 'Variegata', which has cream-splashed foliage; 'Rubrifolia', an introduction of 1878 with large and handsome maroon leaves; and the rose plantain, *P. major* 'Rosularis', which dates back to before the Elizabethan period and instead of the usual flower-spike produces bunches of ruffled green leaf bracts that look like rosettes. And although this book will consider all kinds of plants, it is those that over the centuries have been selected and bred for our gardens that will be my principal focus. That said, when a botanical name is particularly intriguing, I have included it even if it is restricted to, say, a rare species of fungus or has been superseded by something more taxonomically accurate but linguistically impoverished.

It just remains to be said that because many botanical names were formulated in the very distant past, there is sometimes disagreement about both their origins and what they mean. The proliferation of the internet, indispensable as it can be, often confuses matters further, with both experts and amateurs confidently stating as fact what often turns out to be mere conjecture or the duplication of earlier mistakes. Even a plant as familiar as the lupin may owe its name to two entirely different words, one Latin and one Greek, and arguments about the naming of the *Sequoia* have led to long essays by botanists and taxonomists in scholarly magazines. In such cases, I have tended to give both alternative theories rather than come down on one side or the other. When even Royal Horticultural Society publications get things demonstrably wrong, what hope is there for an enthusiastic amateur such as myself? If in doubt, however, I have usually followed W. T. Stearn, even when (as in at least one case) the *Oxford English Dictionary* disagrees with him.

In short, this book is not a work of scholarship, but is intended to inform and entertain. I would also like to think that it will introduce readers to unfamiliar or forgotten plants they may go on to grow in their own gardens. Researching the book has certainly encouraged me to do this. Most books on botanical Latin take the form of dictionaries, whereas this one is intended to be read as a series of discursive essays, to which brief word-lists act as a supplement. It should be evident from its size that it makes no attempt to be comprehensive, and I am cheered by the subtitle of *Stearn's Dictionary of Plant Names for Gardeners*: 'A Handbook on the Origin and Meaning of the Botanical Names of some Cultivated Plants'. 'Some' is all that I am aiming for, and other books with grander aspirations are listed in the bibliography.

ESSENTIALS

Gardeners will be familiar with the basic naming of a plant by genus and species, but there are of course further sub-divisions, as well as rules governing how these should be written, which are useful to learn and are followed in this book. In Britain the proper naming of plants is overseen by the Royal Horticultural Society's Nomenclature and Taxonomy Advisory Group (NATAG), a body of eminent scientists that considers all 'proposals to correct or change names and strives for a balance between the stability of well-known names and botanical and taxonomic correctness according to the codes of nomenclature'. These codes are laid out by the International Commission for the Nomenclature of Cultivated Plants (ICNCP), and are regularly updated.

The basic name of genus and species should be rendered in italics, with the genus name having an upper-case initial, the species name a lower-case one: *Achillea millefolium*. In listings or any writing about plants that refers to more than one species, the genus name is often reduced to its initial after the first mention. For example: 'Many gardeners grow *Achillea millefolium* because of its wide range of colours, from cream to crimson, although the intense yellow *A. filipendulina* and the white, loose-headed *A. ptarmica* are also popular.'

The genus and species names may be followed by that of a cultivar (a contraction of 'cultivated variety'), denoting a different form of the plant, often showing a variation in flower or foliage colour. Unlike the genus and species, which are rendered in Latin and in italics, the cultivar should be in a contemporary language such as English, in roman type, with upper-case initial letters, and placed within single inverted commas: *Achillea millefolium* 'Lilac Beauty'. Since 1959 cultivars 'that are wholly or in part in Latin' have been deemed 'invalid' by the ICNCP code, so out go the previously acceptable 'Variegatum', 'Superba' and (thank goodness) 'Pixie Alba'. Those named before 1959, however, such as the lovely double cranesbill *Geranium pratense* 'Flore Pleno', have been allowed to stand, and you will still find many Latinate cultivar names. The code also states that cultivar names should not be translated from another language, so that *Salvia nemorosa* 'Schneekönig' cannot be renamed 'Snow King' for nurseries in English-speaking countries. The ICNCP also, somewhat snootily, instructs that cultivar names should be differentiated from 'trade designations' (a name 'used to market a plant when the cultivar name is considered unsuitable for selling purposes'): names of this sort should be in roman, of a different typeface from that of the genus and species, and not be placed within inverted commas – something most nurseries happily ignore.

In addition to cultivars there are three further potential subdivisions of a plant species: subspecies, variety and forma. Unlike cultivars, which are deliberately brought into cultivation from sports found in the wild or variant seedlings isolated for this purpose, these three subdivisions are principally botanical. A subspecies is usually a variant that occurs wild in a different geographical location from the species: *Salvia nemorosa* subsp. *pseudosylvestris*. A variety has a different botanical structure from the species: *Salvia nemorosa* var. *fugax*. A forma of a plant denotes some minor botanical difference, such as flower colour: the white foxglove *Digitalis purpurea* f. *albiflora*.

A hybrid form is the result of cross-pollination, occurring either naturally or by human intervention, between genetically different parents. Where human agency is involved, the aim of hybridisation may be to create an aesthetically 'improved' version of the originals, to attempt to stabilise plants that tend to be variable in form or colour, or to make a plant less prone to disease. The correct form for a hybrid is *Salvia* × *sylvestris*.

A final term you occasionally come across but do not need to worry about is group, which gathers together two or more similar cultivars, hybrids with uncertain parentage, or a mixture of species and hybrids sharing characteristics. The group follows the species name and is correctly rendered in roman type, without inverted commas, the name taking an upper-case initial, but 'group' not doing so (though books and nurseries often go their own way on this): *Hydrangea aspera* Villosa group, *Actaea simplex* Atropurpurea group.

Although botanical Latin is different from classical Latin, it observes some of the same basic rules. Most of these need not worry gardeners, but a word or two about gender may be useful. In essence, species names – specific epithets – being adjectives usually match the gender of the genus name, so that they are both masculine, both feminine or both neuter. This sounds alarming, but is in fact relatively easy to grasp. As we have seen, most names, even those derived from the Greek, have Latinised endings; the most common endings for names that have kept their Greek form are also included in the guide-lines opposite and overleaf.

Genus names

Names ending in -*us* or -*er*, e.g. *Cistus, Helleborus, Lathyrus, Cotoneaster*, are masculine.

Greek names ending in -*on*, e.g. *Penstemon*, are usually masculine.

Names ending in -*a*, e.g. *Achillea, Salvia, Pulmonaria*, are feminine.

Greek names ending in -*is*, -*e*, -*ys* or -*odes*, e.g. *Myrrhis, Anemone, Phyllostachys, Omphalodes*, are feminine.

Names ending in -*um*, e.g. *Eryngium, Geranium, Veratrum*, are neuter.

Greek names ending in -*ma*, e.g. *Ceratostigma*, and some ending in -*on*, e.g. *Rhododendron*, are neuter.

Species names (specific epithets)

Masculine forms end in -*us*, -*er*, -*is*: so, for instance

Cistus laurifolius (laurel-leafed cistus)

Lathyrus odoratus (sweet pea)

Helleborus niger (Christmas rose)

Cotoneaster horizontalis (wall spray)

Feminine forms end in -*a*, -*ra*, -*is*: so, for instance

Achillea ptarmica (sneezewort)

Salvia sclarea (clary sage)

Pulmonaria rubra (red lungwort)

Myrrhis odorata (sweet cicely)

Anemone hortensis (broad-leaved anemone)

Phyllostachys nigra (black bamboo)

Neuter forms end in *-um*, *-rum*, *-e*: so, for instance

Eryngium maritimum (sea holly)

Geranium sanguineum (bloody cranesbill)

Veratrum nigrum (black false hellebore)

Rhododendron nivale (snow rhododendron)

Ceratostigma willmottianum (Chinese plumbago)

The usual comparative forms end in *-ior* for masculine and feminine, *-ius* for neuter (as e.g. *elatior/elatius*, meaning 'taller' or 'rather tall'); *major* and *minor* become *majus* and *minus* with neuter genus names.

Some specific epithets remain the same for all genders, among them *praecox*, *splendens* and *tricolor*.

There are of course exceptions.

Most trees take a feminine specific epithet even though many of their genus names look as though they should be masculine or neuter. Other rule-breakers are merely perverse, *Achillea millefolium* being one example.

In some cases the specific epithet is not an adjective but a noun, appearing in the genitive form – in other words 'of' something or someone. This style is particularly used when deriving a specific epithet from a person's name: achieved by adding the ending *-ii* for a man or (rather less often) *-iae* for a woman. *Primula sieboldii* in fact means 'Siebold's primula' and is named after the German botanist Philipp von Siebold (1796–1866), while *Protea roupelliae* means 'Roupell's protea' and is named for the English painter specialising in South African flora, Arabella Elizabeth Roupell (1817–1914). Here again there are exceptions, with sometimes only a single -*i* being added as suffix, as in *Polygonatum hookeri*, named for the explorer and Curator of Kew Botanical Gardens Sir Joseph Dalton Hooker (1817–1911). Another option is to create an adjective from the surname, typically by adding -*ianus*, -*iana* or -*ianum* (since of course the adjective agrees in gender with the genus name), as in the case of Ellen Willmott (1858–1934), the amateur plantswoman who created the now lost garden at Warley Place in Essex, for whom *Ceratostigma willmottianum* is named. Hooker, too, has plants named for him in this fashion: *Elaeocarpus hookerianus* and *Sarcococca hookeriana*.

For those who want to work out the derivation of botanical names for themselves, or simply explore them more, access to a Latin dictionary might prove useful. I'd suggest that the *Pocket Oxford Latin Dictionary* – the pocket perhaps being the largeish one of a gardener's apron – is more than sufficient. You will see that in such dictionaries both the nominative and the genitive case of nouns are given, because this shows the declension of the noun. This is not something gardeners need worry about, but it is a standard principle in both Latin and Greek that the genitive root is used in derivatives or when combining portmanteau words.

There is a problem in transliterating Greek words, with little consistency to be found in botanical dictionaries and other horticultural books, where spelling, accents and diacritic marks often widely differ. Given that, for the purposes of this book, Greek words are instanced merely to show derivations, my not entirely satisfactory solution has been to adopt the simplest form, relying where possible on the *Concise Oxford*

English Dictionary. I have, however, for reasons of clarity, transliterated the Greek letter upsilon as *y* rather than *u*.

Although now 165 years old and restricted to British wild flowers, the Rev. C. A. Johns's *Flowers of the Field* (1853) has a clear and brief Introduction on the structure of plants and the names of their different parts and types. The book went through innumerable editions and is widely and inexpensively available from second-hand dealers, but for choice, and with all due respect to the original author, I'd go for the 29th edition (1899), which was revised and enlarged by G. S. Boulger, professor of botany at the City of London College.

I have taken all illustrations from Gerard's *Herball.* The names of many plants, both common and botanical, have changed since Gerard's day, and most of the captions follow current usage. In some cases, however, it is impossible to match Gerard's plants exactly to those mentioned in my text, and I have therefore used the English names he gave them, retaining his spelling and putting them between single inverted commas.

SPECTRUM

Colour in the garden tends to be a controversial subject, with one person's 'bright and cheerful' being another person's 'garish and tasteless'. In the late nineteenth century both William Robinson and Gertrude Jekyll reacted against the prevailing fashion for regimented Victorian planting schemes in which bedding plants were arranged in geometric patterns. This is a style of gardening that still prevails in some public gardens and at its best it can be very effective, if labour-intensive. In *The English Flower Garden* (1883), however, Robinson argued for a more 'natural' style of gardening, and this was put into practice by Jekyll both in her own garden at Munstead Wood in Surrey and in the gardens she designed for the Arts and Crafts houses of Edwin Lutyens, the architect with whom she frequently collaborated. Rather than following the strict patterns of carpet bedding, Jekyll filled her borders with drifts of flowers, in which one colour gradually merged with another. It has been suggested that this was how she saw the world, her bad eyesight having decided her to give up her first career as a painter and become a garden designer instead. This 'English' style of gardening, shown off to best advantage against large houses of grey or honey-coloured stone, towering yew hedges and elaborate topiary, was hugely influential even for gardens planned on a much smaller scale.

There were nevertheless always gardeners who preferred something a little bolder in colour than the kind of dusty blues, pinks and lilacs particularly favoured by Jekyll, and it is a nice irony that the person who became the most famous and effective advocate for a much more adventurous palette did so at a house that had been remodelled by Lutyens. At Great Dixter in Sussex Christopher Lloyd created borders in which flowers appeared to have been selected for their sheer vibrancy, and introduced colour combinations that most gardeners do their best to avoid. Thinking no one would believe me unless I produced evidence, I once took a photograph of a bed there in which a lipstick-pink fuchsia, an orange impatiens, a bright yellow daisy, a magenta petunia and a salmon-flowered pelargonium with cream-marbled leaves were planted in one retina-searing group. Some gardeners have immovable prejudices against certain colours and the petunia alone would have horrified E. A. Bowles, who greatly liked geraniums but warned that 'the family inherits a pernicious habit of flaunting that awful form of original sin, magenta, and rejoicing in its iniquity'. For all its flaunting, Great Dixter remains one of the great English gardens, but attempts to emulate Lloyd's outrageous colour clashes do not always end happily.

Many plant breeders and nurseries hope to entice their customers by shouting out flower colours in cultivar names, so you know what you are getting if you order pelargoniums called 'Maverick Scarlet', 'Shocking Orange' or 'Custard Cream'. Latin species names, however, can also be helpful guides to colour, and are often discriminating as to shade. It is as well to know that, unless you have the yew-divided acres of Great

Dixter to play with or Christopher Lloyd's maverick eye, a plant with the species name *aurantiacus* (orange) might not be shown to its best advantage planted alongside one identified as *roseus* (pink).

Whatever we may now think of colour, it certainly excited English gardeners in the late sixteenth and early seventeenth century, when it seemed every ship brought new introductions from abroad, and plant collectors went on trips to distant lands to find 'new' species, document them and bring them back to grow in English gardens. The cultivation and study of plants had at last been wrested from the apothecaries and physicians and become a consuming interest for those of means and leisure. Once people started growing plants for pleasure rather than for medicinal use, the flower came into its own and colour became something to be sought out. And indeed sorted out. John Parkinson's *Paradisi in Sole Paradisus Terrestris* may have listed the medicinal 'vertues' of plants, but its main section is titled 'The Garden of pleasant Flowers' and the book opens with advice about how to create such an entity. Chapter IV details 'The nature and names of divers Out-landish flowers, that for their pride, beauty, and earlinesse, are to be planted in Gardens of pleasure for delight'. 'Out-landish' here means flowers from beyond the British Isles, flowers that are 'strangers unto us, and [give] the beauty and bravery of their colours so early before many of our owne bred flowers, the more to entice us to their delight'.

Most of these novelties are now very familiar – 'Daffodils, Fritillarias, Iacinthes [hyacinths], Saffron-flowers [crocuses], Lilies, Flowerdeluces [irises], Tulipas, Anemones, French cowslips, or Bears eares [auriculas]' – but Parkinson gets into considerable difficulties when he comes to list them, often (as Gerard and others did before him) cataloguing the same

plant as several different species because of slight variations
in colour. The wood anemone, *Anemone nemorosa*, is native
to Britain, but the number of other 'winde-flowers' appears to
have been increasing exponentially at this period. '*Dodonaeus*
hath set forth five sorts; *Lobel* eight; *Tabernaemontanus* ten,'
wrote Gerard. 'My selfe have in my garden twelve different
sorts: and yet I do heare of divers more differing very notably
from any of these.' Thirty-two years later, Parkinson confesses
that 'to describe the infinite (as I may so say) variety of the
colours of the flowers, and to give to each his true distinction
and denomination, *Hic labor, hoc opus est*, it farre passeth
my ability I confesse'. Various species of anemone had been
introduced into Britain during the Elizabethan period, mostly
(according to Parkinson) from Constantinople, but also from
Italy, Cyprus and Persia. Parkinson lists several pages of them,

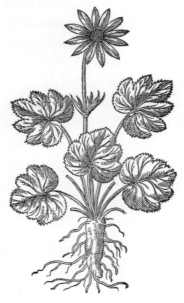

'Broad leaued Winde-floure',
Anemone hortensis

and amongst this floral bounty
is what he calls *Anemone purpurea Stellata altera* or (rather
despairingly) 'Another purple
Starre Anemone' to differentiate it from *Anemone
latifolia purpurea stellata sive
papaveracea*, which is merely
'The purple Starre Anemone
or Windflower'. We now know
that both plants are the broad-
leaved anemone, *A. hortensis*,
which has the synonym *A. stellata*, and that to assign them to
different species was a mistake.
Parkinson goes on to pro-
vide additional Latin names

to differentiate plants within this group that have further variation in flower colour. So the already long Latin names had yet more adjectives added to them: *Persiciviolacea*, 'betweene a Peach colour and a Violet, which is virtually called a Gredeline' (a grey-violet colour more usually spelled gridelin, from the French *gris de lin*, linen-grey); *Cardinalis*, 'of a rich crimson colour'; *Sanguinea*, 'of a deeper, but not so lively a read'; *Cramesina*, 'of an ordinary crimson colour'; *Coccinea*, 'of a Stamell colour, neere unto Scarlet' (stammel was a red woollen fabric used in the medieval period in the manufacture of undergarments); *Incarnata*, 'of a fine delayed red or flesh colour'; *Incarnata Hispanica*, 'of a lively flesh colour, shadowed with yellow'; *Rubescens*, 'of a fair whitish red'. And that is just the reds. *Anemone tenuifolia* (now *A. capensis* var. *tenuifolia*) is equally divided into several species, the names sometimes stretching out towards infinity. One 'whose flower is pure white with blewish purple thrums [threads] in the middle' is bowed down with the name *Anemone tenuifolia coccinea simplex Alba flaminibus purpureis*. These kinds of descriptive names clearly needed shortening and sorting out, and there gradually came to be agreement on single adjectives to describe individual colours. Some of these are more or less the same as those Parkinson used: *cardinalis, sanguineus, carneus* and *coccineus*. (Parkinson's *cramesina*, incidentally, is not a misprint for, say, *carmesina*, and survives as the specific epithet for an uncommon and luridly red-orange fungus that grows in New Zealand, *Dermocybe cramesina*.)

Differentiating between the different shades of red, or indeed any other colour, that occur in plants is difficult because botanical authorities sometimes disagree – even with themselves. In his *Botanical Latin*, for example, William Stearn defers to the terms listed by the German botanist G. W.

Bischoff in his *Handbuch der botanische Terminologie* (1830), reproduced in translation by J. Lindley in his *Introduction to Botany* (1832), and to an article on the subject by B. D. Jackson in the *Journal of Botany* (1899). This means that *puniceus* is defined as 'the purest red, without any admixture'; but in Stearn's later *Dictionary of Plant Names for Gardeners* he defines it as 'reddish-purple'. It is an abiding difficulty with plants named by colour that people down the ages have had different ideas of what constitutes a particular shade. In addition, colour is notoriously difficult to put into words, although some of the great garden writers, such as Reginald Farrer and E. A. Bowles, did so with enormous élan, producing elaborate and often idiosyncratic descriptions that are a joy to read but not necessarily definitive. Bowles, for example, refers to a 'fledgling' of *Geranium endressii* 'with many more spoonfuls of cream than is usual mixed with its raspberries', and 'a curious form of *G. pyrenaicum* that is not quite white, but a faint almost subdued lilac, that reminds me of a red nose smothered in powder'. Farrer famously described an auricula as 'precisely that rich clear colour which is the Imperial yellow of the Chinese court, and which I last saw in the palanquin that was conveying the Empress Dowager to the railway station in Peking'. He nevertheless observed that 'one can only throw one's hands up in despair, confronted at once by the inadequacy of language and by the unending variety and delicacy of flower-tones where no two whites or blues are alike, but all have to be lumped under the one rough heading, eked out by such explanatory qualifications as each separate mind has to hammer out for itself, more in the hope of satisfying itself than of carrying a true picture to others'. It did not much help matters, Farrer thought, that nurseries were inclined to oversell plants by exaggerating the colours of their flowers. Magisterially unimpressed by the kind of fritillaries that many gardeners seek out for their unusual shades, he writes: 'an enormous number of Fritillaries have more or less stinking bells of dingy chocolate and greenish tones, which often appear transfigured by the enthusiasm of those who desire to get rid of them, as "rich purple" or "amaranthine violet"'.

In botanical Latin it is sometimes hard to work out why certain plants have been associated with certain colours or shades, and while specific epithets can be useful they are not always wholly reliable as a guide if you are planning a border. Indeed, some of them are simply baffling, as we shall see. In 1938, however, the British Colour Council in association with the RHS published a two-volume *Horticultural Colour Chart* that attempted to 'fix' the colours of plants. Sometimes known as *The Wilson Colour Chart* (after the then director of the BCC, Robert F. Wilson), this publication had detachable pages so that gardeners could hold the printed colours against flowers for comparison, arriving at a supposedly accurate definition. The chart has since been revised and is now in its 6th edition (2015). It features 920 colours arranged in four fans and resembles a very expensive (£199.00) paint-colour chart. Of course these publications came too late to sort out what constituted *puniceus* or any of the other species names related to colour, the majority of which had been coined many years before.

Given the wide range of colours available to gardeners, it is difficult to know where to start; but if we follow the order of the rainbow, we begin with red, of all the colours the one that, as painters know, has the unfailing ability to draw the eye. Perhaps the most striking plant with the disputed species name *puniceus* is the New Zealand glory pea, *Clianthus puniceus*, which has clusters of flowers in the shape of a parrot's beak that are certainly a very vivid red. More familiar is the pomegranate, *Punica granatum*, the genus name of which derives from the same root. The Romans first imported the pomegranate from Carthage and so called it the Carthaginian apple, *punicum malum*; *punicus*, the Latin term for anything relating to that city state, was derived from the Greek *Phoinix* = Phoenician, the people in whose empire Carthage belonged. Because the pomegranate has bright red flowers *puniceus* subsequently became the Latin word for scarlet or crimson.

Other species names that refer to red at its purest are *carmineus* (from which we get the words carmine and crimson) and *coccineus*, both derived from an insect that provided a highly valuable red dye. These terms may come from entomology, but they reach back through that to botany again and the holly-leafed evergreen kermes oak, *Quercus coccifera*, which grows in Mediterranean regions. The species name for this oak refers to the fact that it is host to a scale insect, *Kermes ilicis*: the Romans called this insect *coccum*, taken from the Greek *kokkos*, the name given to it by Dioscorides, and the suffix *-fera* means 'bearer'. *Kokkos* actually means berry or grain, which the tiny female kermes insect resembles. It is from the females that a vivid red pigment was extracted and used to dye cloth. As with *puniceus*, *coccum* (and its associated adjective *coccineus*) came to mean the colour, and indeed any cloth dyed red. As a specific epithet *coccineus* generally denotes a very bright red, like coccineal, the colourant derived many centuries later from a different scale insect, *Dactylopus coccus*. The vibrant flowers of the scarlet runner bean earned the plant its botanical name of *Phaseolus coccineus*. Plants such as the Texas sage, *Salvia coccinea*, the red passionflower, *Passiflora coccinea*, or the Chilean fire bush, *Embrothrium coccineum*, match this intensity of red, but others, such as the scarlet rose mallow, *Hibiscus coccineus*,

seem rather more variable in colour than the name would suggest. Meanwhile the cultivar name 'Coccineus' has been used
for plants that simply don't make the grade at all: *Centranthus
ruber* 'Coccineus' is perhaps a slightly deeper red than the
species, but the creeping thyme *Thymus praecox* 'Coccineus' is
barely red at all. The similarly derived epithet *kermesinus* is also
unreliable: *Rhododendron* 'Kermesina' signally fails to live up
to its name, its flowers being the kind of deep, unyielding pink
all too common in azaleas, while those of the so-called crimson
passionflower, *Passiflora kermesina*, are the colour of borscht.
The South African cliff lily, *Gladiolus carmineus*, is, however, a
properly bright scarlet.

   Producing scarlet was once an expensive business and the
colour was therefore adopted by those, such as monarchs and
popes, who wished to display their wealth or power. It was first
decreed in 1464 that cardinals in the Roman Catholic Church
should wear scarlet, and this adoption is commemorated in
the specific epithet *cardinalis*. One of the most flaming red
plants in any border is the stately *Lobelia cardinalis*, introduced
from the Americas into Europe in the 1620s during the period
that Pope Urban VIII was in the Vatican. Urban was a ruin-
ously extravagant patron of the arts, and in 1627 had himself
painted by Pietro da Cortona wearing a deep scarlet cape and
cap, seated on a gold and scarlet chair. It seems altogether
appropriate that two synonyms for the plant are *Lobelia fulgens*
(meaning shining) and *L. splendens* (self-explanatory), and
that – adding royalty to the mix – the spectacular purple-leafed
cultivar is named 'Queen Victoria'. Other plants belonging to
the papal conclave include the Mexican sage, *Salvia cardinalis*, and the scarlet monkey-flower, *Mimulus cardinalis* (now
Erythranthe cardinalis), both of them good strong reds, if not
as blazing as the lobelia.

To call a plant *Erythranthe cardinalis* would seem to be over-doing it rather, since the genus name means much the same thing, from the Greek *erythros* (red) + *anthos* (flower). It would be far too simple if any genus name beginning with the prefix *erythr-* designated a plant with red features, in the way that the specific epithets for the red baneberry, *Actaea erythrocarpa* (with red fruit) and the dwarf lady's mantle, *Alchemilla eryth-ropoda* (with red stems) do. While many species of *Erythrina*, such as the Indian coral tree, *E. variegata*, or the Brazilian coral tree, *E. falcata*, do indeed have blazing scarlet flowers, others such as the South American *E. velutina* do not. When Linnaeus named the genus, he seems only to have seen red-flowered species, including *E. variegata* (which he called *E. orientalis*), the coral bean *E. herbacea*, and the coral tree *E. corallodendron*. The flowers of *Erythraea*, whose name means 'reddish', are stub-bornly pink, but the genus is now more commonly known as *Centaurea*. Those of *Erythrochiton brasiliensis*, if white, at least emerge from red bracts, *khiton* being the Greek word for a kind of tunic. And it seems that *Erythronium* was merely adopted from another plant for the dog's-tooth violet, which has flowers that are yellow, white or various shades of pink – but definitely not red. The name originates in the Greek *satyrion erythronion*, meaning a red-flowered orchid and evidently referring to some other plant altogether.

Another Latin word for red is *ruber*, familiar to us in such English words as ruby, rubicund and rubric – this last derived from the fact that headings and instructions in liturgical books are traditionally printed in red ink. In horticulture *ruber* is a generic red and has several variants – *rubens, rubeus, rubescens, rubellus, rubicundus* – as well as forming part of such com-pounds as *rubrifolia* (red-leaved), *rubricaulis* (red-stemmed) and *rubromarginatus* (red-edged, as in the succulent *Echeveria rubromarginata*). *Pulmonaria rubra* is clearly differentiated from other lungworts by its dull red flowers, while the red hook sedge *Uncinia rubra* is named for its leaves, and the redcurrant *Ribes rubrum* for its glistening ruby fruit. At the lighter end of the colour chart are *rubescens* (becoming red, blushing) and *rubellus* (pale red). The latter may be familiar from rubella, the medical name for German measles, so called because it produces a reddish rash. In botany it denotes a purer and unspeckled colour, though sometimes so pale as to be nearer pink, as in *Lilium rubellum* or *Chrysanthemum rubellum*. The epithet *atrorubens* (the prefix taken from the Latin *ater* = black) supposedly refers to plants whose flowers are a very dark red indeed, but this is not always or even often the case. The broad-leaved helleborine, *Epipactis atrorubens*, has flowers of red verging on purple-brown, rather like those of *Geranium phaeum* or *Aquilegia vulgaris* 'Black Barlow', but the darkness of other plants is less easy to spot. *Eupatorium atrorubens* has lilac flowers, but the name refers to the flower-stems and the ribs of its leaves, which are a vibrant maroon. In the case of *Echinacea atrorubens* and *Helianthus atrorubens*, the flowers are respectively purple and yellow, and it is the dark central bosses that give them their botanical names. The madder plant, the roots of which were the source of a flaming orange dye, has the botanical name *Rubia tinctorum*, its species name being the genitive plural of the Latin *tinctor* and so meaning 'dyers' madder'. The flowers of madder are yellow, incidentally, and we tend most often to associate the plant's name with rose madder, a pigment used by artists. Rose madder is also manufactured from *R. tinctorum*, but only after further processing which produces the kind of wonderful vibrant pink that in India is known as *rani* (or 'queen'), a popular colour for both saris and *abir*, the powdered pigment that is flung around during the festival of Holi.

Another red species name related to India and to paint is *miniatus*, a very bright red, like vermilion or cinnabar. Think of the vibrant red markings on the wings of the cinnabar moth (*Tyria jacobaeae*) – though confusingly this insect's own Latin name is derived from Tyrian purple, of which more later. *Miniatus* comes from minium or red lead, a bright pigment that (though toxic) was less expensive than vermilion, which was mercury sulphide extracted from the mineral cinnabar. Minium was favoured by illuminators of medieval manuscripts and later by painters of Indian miniatures. Indeed, the word miniature in this context has nothing to do with the size of the paintings, as most people assume, but derives from *miniatura*, the name given to the work of artists who used minium pigment to paint headings or the initial letters that drew attention to the start of a new paragraph. The most familiar plant named for this colour is a subspecies of the Persian everlasting pea, *Lathyrus rotundifolius* subsp. *miniatus*, which in fact tends to be a pinkish brick-red. It does, however, really come from Persia, a country which had a profound cultural influence on the Mughal Empire, during which Indian miniatures reached their highest perfection. Nearer to the true red of minium is the scarlet passionflower, *Passiflora miniata*, or the Natal lily, *Clivia miniata*, one of two red-flowered plants still sometimes called kaffir lilies, the other being the equally red *Schizostylis coccinea* (syn. *Hesperantha coccinea*). Both plants originate in South Africa, where 'kaffir' is an offensive (and now legally banned) term for a black African, and so the latter plant is now more decently called the crimson flag lily.

There are two specific epithets for blood-red, *sanguineus* and *cruentus*, both of which are Latin words for bloody or gory. The former is more usual, although few of the plants that bear it suggest the operating theatre or butcher's block. The flowering currant *Ribes sanguineum*, once a staple of the cottage garden, has flowers that are more magenta than red, even in such optimistically named cultivars as 'Pulborough Scarlet'. Similarly, coral bells, the common name of *Heuchera sanguinea*, is a better description than the botanical one. The flowers of the red angel's trumpet, *Brugmansia sanguinea*, do at least look as though their mouths have been about vampiric business, while in winter sunlight the new stems of *Cornus sanguinea* glow a beautiful gory red. People are often puzzled by the name of the bloody cranesbill, *Geranium sanguineum*, the flowers of which come in almost every colour except red; but the specific epithet refers to the leaves, which turn a very satisfying crimson in the autumn. While the flowers of the cineraria *Senecio cruentus* are not in the least sanguinary, and those of prince's feather, *Amaranthus cruentus*, tend to purple, those of *Dianthus cruentus* certainly earn it the common name of the bloody pink and effectively differentiate it from the genus's more usual pastel shades. *Atrosanguineus* means dark blood red, perhaps best displayed by the cinquefoil *Potentilla atrosanguinea*, whose shining petals provide a proper splash of arterial red in any border. Other flowers such as those of the once highly fashionable chocolate-scented *Cosmos atrosanguineus* or the small but thrillingly dark *Allium atrosanguineum* tend to the venous.

Although *fulgens* and *fulgidus*, from the Latin verb *fulgere* meaning to gleam, denote shining or brightly coloured, they are usually applied to plants whose flowers are red, like the cardinal lobelia mentioned earlier. Though variable in colour, the long trumpets of *Fuchsia fulgens* (sometimes known as the brilliant fuchsia) can certainly be very red indeed, especially in the cultivar 'Rubra Grandiflora', as are the flowers of *Pelargonium fulgidum* and the succulent *Senecio fulgens*. The coneflower *Rudbeckia fulgida*, however, shines out a bright orange-yellow.

Flame red is denoted by several specific epithets including (from the Latin) *igneus* and *flammeus*, and (from the Greek and far less common) *phlogiflorus*. The glowing little tubular flowers of *Cuphea ignea* certainly suggest, if not a flame, then the burning end of a cigarette, and it is sometimes called the cigar plant. Similarly the flowers of *Rhododendron flammeum* blaze from bright red to orange with a dark mark on the upper lobe that suggests scorching. But the angel's fishing-rod *Dierama igneum* has flowers of a rather grubby pink or lilac and those of the mock verbena *Glandularia phlogiflora* come in red, white, and various shades of purple. How the last named got its name is a mystery equalled by that of the *Phlox* genus, whose name comes from the same Greek root and which has flowers in almost every colour other than flame – though the garden historian Alice M. Coats states that the name was originally used by Theophrastus 'for some other flower of that colour, never since identified'. Thus do botanists make nonsense of their own science.

Although there are many plants with orange flowers, there are very few specific adjectives denoting this colour. Two of them derive from a Greek word we have already come across, *krokos*. The crocus is the source of saffron, and so both these names denote that colour, ranging from the rich yellow of *croceus* to the rich orange of *crocatus*: think of the saffron-dyed robes of Hindu or Buddhist monks, which show the same colour range. The ice poppy *Papaver croceus* (easily confused with the Iceland poppy, *P. nudicaule*) has flowers of a good, strong deep yellow, as does the rock rose *Helianthemum croceum*, and the golden flag, *Iris crocea*, whereas the flame freesia, *Tritonia crocata*, is a dark, fiery orange. That said, as its common name suggests, the coppery ice plant, *Malephora crocea*, is a burnt orange. Another orange is *testaceus*, from the Latin *testa* meaning earthenware. The Nankeen lily, *Lilium* × *testaceum*, is, however a very pale orange-apricot, far removed from ancient potsherds. More reliable is *aurantius* or *aurantiacus*, both derived from *aurantium*, the Latin word for an orange tree – the colour was named after the fruit rather than vice versa. The orange tree originated in China and so its botanical name is *Citrus sinensis*, but the darker-peeled Seville orange is *Citrus* × *aurantium*. The orange hawkweed, which most of us know as *Hieracium aurantiacum*, has undergone a genus change and is now *Pilosella aurantiaca*, but still has strikingly orange flowers which give it the alternative name of fox-and-cubs. Various shades of orange are found in the large bottlebrush flower-spikes of the ginger lily *Hedychium aurantiacum*, the fanfare of marmalade trumpets flourished by the jasmine *Cestrum aurantiacum*, and the unusual passionflowers of *Passiflora aurantia*; but one of the best orange-coloured plants is the horned poppy *Glaucium flavum* var. *aurantiacum*, the frail-looking but in fact very robust flowers of which contrast beautifully with the glaucous foliage that gives it its genus name.

The ordinary horned poppy has yellow flowers, *flavus* being one of the specific epithets for that colour, the other being *luteus*, both of them simply Latin words for yellow. From *flavus*, which suggests the purest yellow, we also get *flavens*, *flavidus*, *flaveolus*, *flavicans* and *flavescens*, all denoting different shades of the colour but roughly meaning yellowish. The yellow balsam, *Croton flavens*, has very pale yellow buddleia-like flowers, and the Chinese Ku-shen, *Sophora flavescens*, has flowers of a similar shade, whereas those of the yellow kangaroo's paw, *Anigozanthos flavidus*, are a stronger colour. *Luteus* is often used to distinguish yellow-flowered species of plants that more usually bear flowers of other colours, including the foxglove *Digitalis lutea* and the European lupin *Lupinus luteus.* As with *flavescens*, the specific epithet *lutescens* denotes a paler shade, as can be seen by comparing the primrose-flowered *Rhododendron lutescens* with the much darker and brighter yellow azalea, *R. luteum.* A similar pallor is denoted by *luteolus*, as in dyer's rocket, *Reseda luteola*, or the Australian *Crinum luteolum.*

Other shades of yellow are denoted by *citrinus* (lemon yellow), *cerinus* (wax yellow), *sulphureus* (sulphur yellow), *vitellinus* (egg yellow), and *stramineus* (straw yellow). Rather less alluring are *squalidus* and *isabellinus*, both of which denote a grubby yellow. The former seems a rather unfair epithet for the New Zealand daisy *Leptinella squalida*, which is more happily known as brass buttons. The latter is related to the rarely used English adjective isabel, meaning 'a yellowish-grey or drab colour' and now largely remembered for the famous dismissal of a false etymology in the *Chambers Twentieth Century Dictionary* (1901): 'Not from *Isabella*, daughter of Philip II, wife of the Archduke Albert, who did not change her linen for three years till Ostend was taken.' Whatever the derivation, the foxtail lily *Eremurus* × *isabellinus* has flowers ranging widely in colour through the yellow-pink spectrum, though those of the cultivar 'Isobel' [sic] are indeed a pale (if not especially dingy) yellow.

A Greek word for yellow is *xanthos*, from which we get *xanthinus*, familiar in the species name of the 'Canary Bird' rose, *Rosa xanthina*. As a prefix *xanth-* is used to denote such things as yellow fruit (the yellow-berried nightshade, *Solanum xanthocarpum*, and the yellow guelder rose, *Viburnum opulus* 'Xanthocarpum'), or yellow bell-shaped flowers (*Rhododendron cinnabarinum* subsp. *xanthocodon*). The epithet *caloxanthus* denotes a yellow deemed particularly beautiful, as demonstrated by the flowers of *Rhododendron caloxanthum.*

The Greek word *khrysos*, meaning gold or yellow, supplies the prefixes *chrys-* and *chryso-*, chrysanthemum simply meaning 'golden flower', and *Crocus chrysanthus* a yellow-flowered crocus. People are sometimes bemused by the botanical name of the delicate flag *Iris chrysographes*, the most striking feature of which is its black flowers. If you look closely, however, you will see faint gold markings, *graphe* being the Greek for writing. The Latin for golden is *aureus*, a word widely used in cultivars with yellow foliage such as *Philadelphus coronarius* 'Aureus' or the golden hop, *Humulus lupulus* 'Aureus'. It is occasionally used as a specific epithet: the golden currant, related to the blackcurrant and redcurrant, is *Ribes aureum*, although this is because of its yellow flowers rather than its amber fruit. The prefix *aureo-* produces *aureomarginatus* (golden-bordered) and *aureomaculatus* (golden-spotted), again more usually found in cultivar names such as *Daphne odora* 'Aureomarginata' and the so-called leopard plant *Ligularia tussilaginea* 'Aureo-maculata'. (Cultivar names sometimes use hyphenated or differently spelled versions of specific epithets.) A final touch of gold is provided by *auratus*, meaning decorated with the colour, as in the yellow stripes on the white flowers of the golden-rayed lily, *Lilium auratum*.

And before we leave the yellow, we should mention *pallidus* and *pallidiflorus*, both of which technically mean pale-flowered but often denote a pale yellow, as in the lemon bottlebrush, *Callistemon pallidus*, the Siberian *Fritillaria pallidiflora* and the primrose-trumpeted *Narcissus pallidiflorus*. There are, however, plants with flowers that are pallid but not yellow, such as *Iris pallida* (pale lilac) and the giant pineapple flower, *Eucomis pallidiflora* (greenish white).

The colour green is denoted by numerous specific epithets and prefixes, a familiar one to gardeners being taken from a

Greek word for the colour, *khloros*, which added to the word for leaf gives us chlorophyll, that essential green pigment that allows plants to absorb energy from sunlight. Given that the foliage of the vast majority of plants is green, specific epithets referring to this colour tend to be used to describe other aspects: thus *chloranthus* means with green flowers, much favoured by gardeners in search of novelties such as *Fritillaria chlorantha* and the orchid *Cymbidium chloranthum*. The number of compound epithets derived from the Greek prefix *chlor-* alone seem to go on for ever. They are not all currently used, but include the colours *chloridius* and *chloroticus* (pale green), *chloromelas* and *melanochlorus* (very dark green), *chlorinus* (yellow-green), *chlorocyanus* (bluish green), *chlorochrysus* (golden green) and *chloroleucus* (greenish white). Other attributes include *chloracanthus* (with green thorns), *chloracrus* (with green points), *chlorocephalus* (with a green head), *chlorochilus* (with green lip), *chlorolomus* (green bordered), *chloronemus* (with green threads), *chlorophanus* (with a green sheen), *chlorophthalmus* (with a green eye), *chlorostachys* (green-spiked), *chlorotrichus* (with green hairs) and *chlorurus* (with green tails).

This would have pleased the classical scholar and gardener John Raven (son of the biographer of John Ray and other early naturalists), who had a particular penchant for green flowers. One of the most popular plants with green-flowered species is the hellebore. The very greenest is *Helleborus viridis*, whose specific epithet is one of two Latin words for green, the other being *virens*. Less usual than a green hellebore is the green lavender, *Lavandula viridis*, which in form resembles *L. stoechas*, with little chartreuse 'ears' on top of its green flower-heads. It also has bright green foliage and a strong, distinctive smell that nurseries like to describe as lemony but to me seems more resinous and nearer camphor. The name of *Salvia viridis* is a puzzle, given that there is nothing especially green about the plant; but it has been proposed (to me rather unconvincingly) that Linnaeus used the epithet because the plant is annual and greenness suggests youth and energy. The RHS calls it green-topped sage, and says the bracts for which people grow the plant are 'green or purple', but cultivars such as 'Alba', 'Oxford Blue' and 'Rose Bouquet' tend to be more popular with gardeners. The epithet *virens* is less often used, and *Penstemon virens*, known in its native America as blue mist beardtongue, presumably owes its species name to the mat of dark green leaves from which the blue flowers emerge. The green bottle-brush, however, is *Melaleuca flavovirens*, denoting a flower of a yellowish green.

More purely green flowers are also designated by *viridiflorus*, and a good deal more reliably than *viridis* in the cases of *Galtonia viridiflora*, with its towering steeples of green bells, *Aquilegia viridiflora*, which strikingly and variably combines pale green with chocolate brown, and even *Narcissus viridiflorus*, a frankly etiolated but undeniably green-tinged alpine species. The startling blooms of *Ixia viridiflora* would more properly be described as turquoise, a shade so rare in flowers that it does not have its own specific epithet. Meanwhile, *viridissimus* supposedly means very green, but I'm not sure anyone looking at *Forsythia viridissima* would be able to tell how it gained this name. The flowers are as unyieldingly yellow as in any other species, but apparently you need to look at the stems.

Given that blue is one of the most common colours for flowers, it is perhaps surprising that it has not given rise to a greater number of specific epithets than it has. The explanation may be that when confronted with so many blue flowers botanists needed to look elsewhere in order to differentiate species. Think of the intense blue of *Salvia patens* or the Himalayan poppy *Meconopsis betonicifolia,* or the range of blues found among eryngiums, geraniums, scabious or agapanthus: none of these have species names that refer to their colour. Even so, there are several specific epithets meaning blue. The Latin words for the colour are *caeruleus* and *cyaneus,* familiar respectively in their English versions to painters (cerulean) and anyone who has a computer printer (cyan). Closely related to the first blue are botanical adjectives derived from the Latin words for the sky or the heavens, *caelum,* and for heavenly, *caelestis.* In their botanical forms both *caeruleus* and *cyanus* or *cyaneus* usually denote the clear, bright blue of the cornflower, *Centaurea cyanus.* Despite numerous pink, white and purple cultivars, the blue of the true cornflower is recognised by its French name, *bleuet,* which also became the French equivalent of 'Tommy' in the First World War because of the colour of the trousers worn by conscripted private soldiers. Like the Flanders poppy, the *bleuet* flourished in the churned-up earth of the battlefields and became a symbol of remembrance. The flowers of Jacob's ladder, *Polemonium caeruleum,* should be a good blue, though most people grow cultivars that – apart from the excellent 'Blue Pearl' – tend towards purple.

Other plants with these sky-blue specific epithets may not look particularly blue, but are more blue than other species in the genus. The blue passionflower, for example, is named *Passiflora caerulea* for the distinctive filaments of the corona, which may look purple but are less so than those of such similar species as the one that produces the best fruit, *P. edulis.* Similarly the blossoms of the blue mist flower, *Conoclinium coelestinum,* may look nearer mauve, but the plant was formerly known as *Eupatorium coelestinum,* and its flowers certainly look blue compared with the pinkish mauve or white of other eupatoriums.

Less ambiguously blue is *azureus*, from the medieval Latin *azzurum* and ultimately derived from *lazward*, the Persian word for lapis lazuli. The Italian bugloss, *Anchusa azurea*, has flowers of a very strong blue, though those of the comfrey *Symphytum azureum* are perversely a less unadulterated blue than those of *S. asperum*, which (along with its height) just about compensate for its rash-inducing foliage. Ultramarine is a pigment extracted from lapis lazuli, its name coming from the Latin *ultra* (beyond) + *mare* (the sea). The story goes that it got its name among painters in sixteenth-century Italy as a pigment that had to be imported from beyond the seas, though it has also been suggested that the colour itself is beyond even the dark blue dazzle of the ocean. Like *azureus*, the specific epithet for ultramarine, *lazulinus*, takes its name from the source of the pigment, hence the dark blue grape hyacinth *Muscari lazulinum*.

The Latin word *caesius* denotes a greyish blue, as in the dusty blue *Allium caesium*: this apparent colour, when seen from a distance, resulting from the small spherical flower-heads having dark blue markings on the pale blue florets. A comparison between this allium and the more clearly blue *A. cyaneum* shows the difference in shade between the two species names. As its common name suggests, the purple mountain bugler has flowers of a light lilac colour, and its botanical name, *Penstemon caesius*, refers to its foliage, which is glaucous bordering on verdigris. The actual Latin word for verdigris is *aerugo*, which also means malice, envy or ill will, from the poetic notion of a mind corrupted and discoloured in the manner of oxidised copper. Once again, the botanical epithet derived from it usually applies to foliage, beautifully exem-plified by the striking and unusual leaves of *Rhododendron campanulatum* subsp. *aeruginosum*. Numerous plants have that grey-blue-green colour we call glaucous, from the Greek word for it, *glaukos*, which becomes the epithet *glaucus*, often used for grasses of that colour, such as the blue sedge, *Carex glauca*, and the blue fescue, *Festuca glauca*. Other species or subspecies marked out by this foliage colour are *Euphorbia glauca* and *Coronilla valentina* subsp. *glauca*, which has the same contrast between bluish foliage and yellow flowers as rue. It is lead (Latin: *plumbum*) rather than copper that gives the *Plumbago* genus its name because of the lead-blue flowers.

Between blue and violet in the rainbow comes indigo, derived from the Greek *indikon*, meaning from India. In fact the indigo plant, *Indigofera tinctoria*, was used to produce a blue dye in many ancient cultures, rather as the most famous fabric associated with it, denim, is no longer the province of Nîmes, the French city that gave it its name. That said, from the last quarter of the eighteenth century the eastern Indian state of Bengal became one of the principal growers of the indigo plant, with farmers persuaded by the British to grow it instead of food crops, a disastrous move which eventually gave rise to the violently suppressed Indigo Revolt of 1859, two years after the so-called Indian Mutiny. *Indigofera* means indigo-bearing, and the blue dye is made by fermenting the leaves of the plant, the pea-like flowers of which in most species are a pinkish purple. Gardeners are less likely to grow dyer's indigo than more decorative species such as *I. heterantha*, or *I. dielsiana*, which is named for the German botanist and collector Ludwig Diels (1874–1945) and has slender spikes of pale pink flowers. The woad with which warring Ancient Britons daubed themselves is the product of *Isatis tinctoria*, which has the synonym *I. indigotica*, but despite its yellow flowers this is unlikely to be grown in the garden of anyone who isn't also interested in craft dyeing.

Just as indigo provides a rather indistinct boundary between blue and violet in the rainbow, so flower colours in this range are sometimes hard to categorise. In Stearn's *Botanical Latin* blue, violet and purple are simply lumped together, and the famed Tyrian purple produced from the glands of the shellfish *Murex brandaris* and *Thais haemastoma* and used to dye the robes of Roman emperors may in fact have been closer to a dark red. The Latin word for violet – both the colour and the plant – is *viola*, hence the genus name *Viola*, which includes the sweet violet, *V. odorata*, and the dog violet, *V. canis*, which are indeed that colour, but also garden violas and pansies, which can be almost any colour, and the wild pansy sometimes known as heartsease, *V. tricolor*, the name meaning three-colour (violet, a darker purple and yellow). The principal specific epithets for the colour violet are *violaceus* and *violascens*. The former is perhaps most familiar as a cultivar name, for the violet-flowered double cranesbill *Geranium pratense* 'Plenum Violaceum' and for the viticella clematis, *Clematis* 'Venosa Violacea'; but there are numerous *violaceus* species, such as *Tulbaghia violacea*, a South African bulb that has been lumbered with the ludicrous English name of society garlic, presumably because the plant is showy but a member of the allium family and has leaves that smell like ransoms. Despite this, I grow the cultivar 'Variegata' because the pretty little flowers show up particularly well against the greyish, white-margined leaves. The violet-streaked new stems of *Phyllostachys violascens* give this bamboo its botanical name. Plant species with dark violet flowers take the usual prefix of *atro-*, as in *Gladiolus atroviolaceus*, a wonderfully dark alternative to the more popular but stridently magenta *G. byzantinus*. The delicate purple mullein *Verbascum atroviolaceum* is also a properly dark colour. A synonym for this plant is *Verbascum phoeniceum*, which takes us back to Carthage and the instability of colour terms derived from ancient cultures, since the specific epithet *phoeniceus* can also denote a red colour and is sometimes regarded as a synonym for *puniceus*. The passionflower *Passiflora phoenicea* happily combines both colours, with bright purple filaments emerging from dark red petals. The Australian bottle-brush plant, *Melaleuca phoenicea*, however, is a very definite red, as is the midday flower *Pentapetes phoenicea*, an unusually scarlet member of the hibiscus family.

Other epithets meaning deep violet include *amethystinus* and *hyacinthinus*. The former is taken from the Latin word for the precious stone, which itself is derived from the Greek *amethystos*, meaning 'not drunken' – because it was believed that alcohol served in goblets carved from this material could be consumed without intoxication. True *Allium amethystinum* is a dark violet, but for some reason people seem to prefer the drab maroon cultivar 'Red Mohican'. The flowers of the shooting star *Dodecatheon amethystinum* are in truth magenta, but the sea holly *Eryngium amethystinum* has flowers, bracts and stalks that are all a spectacular deep violet. The colour hyacinth takes its name from the flowers of *Hyacinthus orientalis*, which in its original form are a dark violet similar to that of the bluebell, *Hyacinthoides non-scripta*. (See also 'Eponyms'.) Triteleia is a much underrated summer-flowering bulb which has grass-like foliage from which rises a short stalk ending in an umbel of six-petalled flowers. I started with *Triteleia ixiodes*, which has flowers of a subtle and unusual straw colour, and would be more usefully known as *'T. straminea'*. I then discovered *T. hyacinthina*, which is more delicate in appearance, but has flowers of the deepest violet.

Another Latin word for violet-coloured is *ianthinus*, variously designated as blue-violet, deep violet or bluish purple.

Would that all plants with this species name were as unequi-vocally and vibrantly violet as the Japanese sea slug *Pteraeolidia ianthina*. The orchid *Dendrobium ianthinum* has pale lilac petals and a deep violet lip and *Eupatorium ianthinum* has heads of tasselled flowers like cheer-leaders' pompoms that almost glow. The latter has been reallocated to a new genus, however, and its species name unfairly downgraded, so that it is now *Bartlettina sordida*, the specific epithet meaning just what it sounds like: grubby.

Botanically speaking, the difference between violet and lilac is in the eye of the namer. Technically a pinkish violet, the colour lilac is best defined in the garden by the shrub of that name, *Syringa vulgaris*. The dark purple buds open much paler, and this is the colour that gives the plant its common name. (The Latin name, devised by Linnaeus from *syrinx*, the Greek word for pipe, refers to the plant's hollow stems, which were used for making music; incidentally, we now use 'syrinx' not only for a set of pan pipes but for that well-developed part of the larynx in birds which produces their singing.) Confusingly, the word 'lilac' is a corruption of the Persian word *nilak*, mean-ing bluish, *nil* being the word for blue not only in Persian but in several other Indo-Aryan languages, such as Bengali. With the specific epithet *lilacinus* we are on rather firmer ground, *Verbena lilacina* displaying the same dark buds and paler flowers as the lilac. *Limonium lilacinum* has unambiguously lilac flowers, but confuses matters with its common name of sea lavender, lavender being a mauve-tinged pale blue. Lavender, however, is a colour that although frequently found in the garden does not have its own Latin epithet. It is generally thought that the plant from which the colour takes its name, *Lavandula*, is itself derived from the Latin word *lavare* meaning to wash, since this highly aromatic shrub was supposedly steeped in water by the Romans to provide a fragrant bath. It has also been suggested, however, that the name comes from the altogether different Latin word *livor*, which means a bluish tinge such as one caused on the skin by bruising.

The Latin word for purple is *purpureus*, which also provides the specific epithet; the Greek is *porphyreos*, which has been Latinised to *porphyreus*. The former is widely used, from the common foxglove, *Digitalis purpurea*, and purple toadflax, *Linaria purpurea*, to the purple

Spike lavender, *Lavandula latifolia*

coneflower, *Echinacea purpurea*, and such hybrids and cultivars as the purple rock rose *Cistus* × *purpureus* and purple fennel, *Foeniculum vulgare* 'Purpureum'. Plants with flowers that differ in colour from those that occur naturally sometimes include two colours in their names: so the purple-flowered dittany is *Dictamnus albus* var. *purpureus* and the white foxglove is *Digitalis purpurea* f. *albiflora*. *Porphyreus* sounds prettier than *purpureus* if one thinks of porphyry, but not if one thinks of porphyria, the medical condition that caused the 'madness' of George III, one symptom of which was that his urine turned the colour of methylated spirits. Most euonymus are grown for their foliage or their brightly coloured fruit, but *Euonymus porphyreus* owes its name to its dark purple flowers. More familiar is the purple elder, *Sambucus nigra* f. *porphyrophylla*. The species name *nigra* refers to the shrub's glistening black

berries, a rich source of Vitamin C, but this increasingly popular forma has very dark purple leaves and pink rather than
creamy white flowers.

So dark – almost black – is the foliage of this elder that
it might well have been named *atropurpurea*. As with other
colours, *atro-* emphasises darkness, as in the vinous Balkan
knapweed *Centaurea atropurpurea* and the velvety *Scabiosa
atropurpurea*, sometimes known as the mourning-bride scabious and available in a number of cultivars with names – 'Chile
Black', 'Black Knight', 'Black Cat' – that emphasise its depth
of colour. The flower spikes of the loosestrife *Lysimachia atropurpurea* are very dark, but as the individual flowers open they
become the colour of damson fool, whereas the wine-dark
foliage of *Berberis thunbergii* f. *atropurpurea* deepens in the
autumn, forming a perfect foil for the red berries. The specific epithet is also used for many dark cultivars, such as the
purple periwinkle *Vinca minor* 'Atropurpurea', which snakes
its dusky way among the hellebores and green heuchera in my
garden, or the handsome burgundy thistle *Cirsium rivulare*
'Atropurpureum'.

At the lighter end of the purples is mauve, a colour only
discovered in 1856 when a teenage student of science called
William Perkin, working in a laboratory he had set up at his
home in the East End of London, was attempting to synthesise
quinine from coal tar. He had no luck producing quinine, but
a by-product of his experiments was a colourfast purple dye
which was at first called aniline purple but became known as
Perkin's mauve. Before Perkin's accidental discovery, all dyes
were made from natural substances and tended to be variable
in colour and to fade in light or with washing. It was a nice gesture that the first synthetic dye should look back to its naturally
produced predecessors, the word mauve being derived from

Malva, the botanical name of the mallow. The flowers of plants in the *Malvaceae* family range in colour from the very pale pink of the marsh mallow, *Althaea officinalis*, to the dark magenta of the purple poppy-mallow, *Callirhoe involucrata*. It was, however, the deep pink-purple flowers of the common mallow, *Malva sylvestris*, that suggested the new colour's name, and the epithet *malvinus* is used for plants with mauve flowers, such as *Plectranthus malvinus*. (The epithet *mal-vaceus*, incidentally, simply means mallow-like and does not refer to colour.)

Marsh mallow, *Althaea officinalis*

We have reached the rainbow's end, without considering one of the garden's most popular colours: pink. As with mauve, heliotrope and lavender, several shades of pink take their names from flowers: rose, fuchsia, amaranth. This last, which is a kind of raspberry pink, was commonly used in the eighteenth and nineteenth centuries but has since fallen into disuse, rather as the plant itself in its most familiar form of love-lies-bleeding, *Amaranthus caudatus*, has fallen out of fashion. The word pink itself appears to be of seventeenth-century origin and derives from *Dianthus plumarius*, the common pink. How the flower got its English name is disputed. The *OED* suggests it is a contraction of 'pink eye', meaning 'a small or half-shut eye', pre-sumably referring to the diminutive but often brightly coloured 'eye' at the flower's centre. ('Pink' in this sense derives from the Dutch word for small – as in the pinkie finger.) Another suggestion is that the plant was named because of the neatly snipped appearance of its petal-edges, as if they had been cut by pinking shears, which are used to make the zigzag edge of cloth that prevents it from fraying – seen most often in fabric samples.

In many languages the word for pink is more or less the same as that for the rose: French *rose*, German *rosen*, Italian *rosa*, Bulgarian *rozov*, Bengali *golapi*, Hindi *gulabi*. It therefore seems appropriate that the most common specific epithet for pink is *roseus* from *rosa*, the Latin word for the plant. An early introduction to British gardens, the common hollyhock is *Alcea rosea*, although even in Gerard's time its flowers were 'sometimes white or red, now and then of a deepe purple colour, varying diversely, as Nature list to play with it'. Parkinson, who called the plant *Malva hortensis rosea simplex & multiplex diversorum colorum*, concurred: 'the flowers are of divers colours, both single and double, as pure white, and pale blush, almost like white, and more blush, fresh and lively of a Rose colour, Scarlet, and a deep red like a crimson, and of a darke red like blacke bloud'. It will be noted, however, that all these colours are in the white-pink-red-purple range. In contrast, *Allium roseum* is a good clear pink, as is the shrub *Deutzia* × *rosea*, whereas other species and hybrids such as the rose periwinkle, *Catharanthus roseus*, the

'Double purple Hollihocke', *Alcea rosea*

Himalayan meadow primrose, *Primula rosea*, and the pink valerian *Centranthus ruber* 'Roseus' are of a deep pink, nearer to fuchsia than to rose.

Another epithet denoting pink flowers is *carneus*, from *caro* (gen. *carnis*), the Latin word for flesh. This sounds more gruesome than it is, although our forebears were a good deal less squeamish than we are when, for example, they used the word horseflesh for the deep maroon colour seen in such flowers as those of the oriental poppy 'Patty's Plum'. The kind of flesh botanists had in mind for *carneus*, however, appears to have been that of a rosy-cheeked Briton and the specific epithet is in fact barely distinguishable from *roseus*. *Polemonium carneum* is the kind of very delicate pink of a young girl's face in a sentimental Victorian portrait, but there is nothing in the least fleshlike about the insistent pink of both *Gladiolus carneus* and the heather *Erica carnea*.

The Greek word for a rose is *rhodon*, and *rhododendros* (literally 'rose tree') was what the Greeks called the oleander. This name was still current in the sixteenth century when Cesalpino wrote that a shrub we can identify as *Rhododendron ferrugineum* had flowers so similar to the 'rhododendron' (by which he meant oleander) that it was sometimes called the 'Alpine Rhododendron'. No one looking at this plant today would think its flowers much resembled those of the oleander, but they are a good deep rose colour. Linnaeus took the Greek word and, as *Rhododendron*, applied it to the shrubs we know by that name today, at the same time separating them out from azaleas. Like hollyhocks, rhododendrons have a wide range of colours, some already mentioned, but in his *Species Plantarum* Linnaeus listed only five species – *R. ferrugineum*, *R. dauricum*, *R. hirsutum*, *R. chamaecistus* and *R. maximum* – and all of them have pink flowers. The prefixes *rhod-* and *rhodo-* can mean red as well as rose-coloured, and the epithet *rhodanthus* is applied to plants with flowers as scarlet as *Hibiscus rhodanthus* or as crimson as the rose mallee, *Eucalyptus rhodantha*. The Australian sun-ray daisy, *Rhodanthe*, however, is usually a clear pink, and as if to settle the question the pink flowers of the queen's crown stonecrop make it *Rhodiola rhodantha* – perhaps to differentiate it from the king's crown, which, although called *R. rosea*, has yellow flowers.

It is with some relief that we turn to white, which is surely an unambiguous colour. One Latin word for white is *albus*, familiar from albino and Albion, the old name for England derived from its white cliffs at Dover. The familiar white water-lily is *Nymphaea alba*, the white false hellebore *Veratrum album*, and the white poplar, so called because of its white bark and downy shoots, *Populus alba*. 'Albus', 'Alba' and 'Album' are also widely used in cultivar names for white varieties of plants that usually have different-coloured flowers, such as the white broom *Cytisus* 'Albus', the white lavender *Lavandula* × *intermedia* 'Alba' and the white monkshood *Aconitum* × *cammarum* 'Grandiflorum Album'. An alternative but less certain epithet is *albidus*, which can mean white-flowered, as in *Penstemon albidus*, but also 'white-leaved' (more truly grey-leaved), as in the rock rose *Cistus albidus*, the flowers of which are a very vivid lilac-pink. The other Latin word for white is *candidus*, which as a specific epithet means a pure shining white suitable for the Virgin Mary, as in the Madonna lily, *Lilium candidum*, and the false mallow *Sidalcea candida*. Other kinds of white include *niveus* and *nivalis* (the Latin words for snowy), as in the white *Crocus niveus* or the common snowdrop *Galanthus nivalis*. The latter epithet can also mean growing near the snowline, and so the alpine *Phlox nivalis* 'Nivea' does not have as tautologous a name as it seems: the specific epithet refers to its habitat, the cultivar name differentiates it from the species, which has purple flowers. Similarly the dwarf snow rhododendron, *R. nivale*, has flowers that range from magenta to pink but grows in the snowy Himalayas. Plants with milk-white flowers are designated *lactiflora*, from *lacteus*, the easily recognisable Latin word for 'milky'. The mugwort *Artemisia lactiflora* has flowers that suggest the creamiest of milk, whereas those of *Campanula lactiflora* are milky blue, though the plant (a late introduction of 1814) is possibly more familiar in such cultivars as the violet 'Pritchard's Variety' or the pink-tinged 'Loddon Anna' – or indeed the pure white 'Alba'.

The Greek word for white is *leukos*, familiar to us in such words as leucocytes or white blood cells. The most common epithet is *leucanthus*, meaning white-flowered. Brooms appear to spend much of their time shilly-shallying between genera,

'Steeple milky Bell-floure',
Campanula lactiflora

so that *Cytisus leucanthus* is now *Chamaecytisus albus*, a name change which at least still lets you know the colour of its flowers. That said, another white-flowered broom dubbed *Chamaecytisus albus* by a different botanist is apparently now *Cytisus multiflorus* – or at any rate it was at the time of writing. *Cephalaria leucantha* is similar to *C. gigantea,* but smaller and with much paler yellow or white pincushion flowers. The white-flowered subspecies of the usually lilac *Crocus goulimyi* is labelled subsp. *leucanthus,* and the moonbeam lily is *Lilium leucanthum* var. *centifolium*. There is also a genus of white daisies, *Leucanthemum*, among which you will find both the ox-eye daisy, *L. vulgare* and the shasta daisy, *L. × superbum*.

In botanical names for colours, grey and silver are more or less interchangeable and almost always refer to foliage. The Latin word for silver is *argentum*, which gives us the specific epithet *argenteus*, which means a particularly lustrous grey, as in the large silver-furred leaves of *Salvia argentea*, which has pure white flowers. The rock rose *Cistus × argenteus* has grey-green leaves that form a perfect foil for its usually pink flowers, especially in the cultivar 'Silver Pink'. Dense hairs on its leaves give the silver tree both its English name and its botanical one, *Leucadendron argenteum*, and silvery foliage distinguishes the purple-flowered *Lupinus argenteus* from other lupins. Cultivars often make use of the Latin word for silver, particularly when foliage is variegated or has silver margins. Variegated privet is *Ligustrum ovalifolium* 'Argenteum' and the holly that I have growing (very slowly) in a large pot is *Ilex aquifolium* 'Argentea Marginata'. The margins of the latter are perhaps nearer a cream colour, but are definitely more silver than some of the unforgiving yellow variegation you get in other hollies masquerading as 'Aurea' i.e. golden.

The Latin word for grey is *cinereus*, from *cinis* meaning ash – hence cinders and Cinderella. This gives us three specific epithets for ash grey: *cinereus* itself, *cinerarius* and *cinerescens*. The grey-green foliage of the ash-coloured speedwell, for example, makes it *Veronica cinerea*. Of the many plants known as dusty miller because of their grey foliage, the knapweed *Centaurea cineraria* subsp. *cineraria* has the advantage of purple thistle-like flowers rather than the coarse yellow daisies of the species, which most gardeners simply snip off. The even less attractive ginger-centred yellow daisies of *Senecio cinarescens* usually meet a similar fate, since the plant is sensibly grown for its blue-grey lobed leaves. Another kind of blue-grey is *lividus*, the Latin word for slate grey or bruise grey, beautifully exemplified by the foliage of *Helleborus lividus*, which has the bluish luminescent sheen of decaying flesh. Defining a more straightforwardly greyish foliage is *incanus*, meaning hoary and derived from *canus*, the Latin word for 'white with age'. *Santolina incana* is a synonym for *Santolina chamaecyparissus* or cotton-lavender, another grey-foliaged plant with yellow flowers that are often removed. By contrast, it is for its headily scented flowers that the Brompton stock is grown, but it owes its botanical name, *Matthiola incana*, to its grey-green leaves. The English name of the pink dandelion draws attention to its unusually coloured flower, but botanists were looking at its greyish foliage when they named it *Crepis incana*. The white or silver horehound, *Marrubium incanum*, has leaves that are not only grey but felted white on the undersides, while it is the tiny white hairs on the stems of *Scutellaria incana* that make it the downy or hoary skullcap.

Flowers that we describe as brown tend in fact to be tawny, properly defined as either orange-brown or yellow-brown. The Latin word for tawny is *fulvus* or (less commonly) *fulvidus*. The former is used reasonably enough for the brownish orange day lily *Hemerocallis fulva* but less so for *Digitalis × fulva*, which is sometimes called the strawberry foxglove, suggesting a colour some considerable way removed from tawny. Also called *D. mertonensis*, this plant has flowers that might more accurately be described as the colour of (proper, home-made) strawberry ice-cream; but other foxgloves do approach brown in colour if not in name. Both *D. lanata* and *D. laevigata* have brownish flowers but take their species name from their respectively woolly and smooth leaves. Better still is D. *parviflora*, but its name refers to the size of the flowers rather than their milk-chocolate colour. Best of all, perhaps, is *D. ferruginea*, the species name taken from *ferrugo*, the Latin word for iron rust, and for once acknowledging the flower colour. We have already mentioned *Rhododendron ferrugineum*, and this got its name from the rusty spots on the undersides of its leaves, whereas the leaves of *Pittosporum ferrugineum* are covered in short, rust-coloured hairs.

I recall being mocked by friends for attempting once to grow a brown-flowered clematis. *Clematis fusca* takes its name from the Latin word for swarthy, and its small urn-shaped flowers are deep chocolate and covered with fine hairs. The interior, should one be able to see it (the flowers don't really open), is white, and a white margin is just discernible on the petals. For the faint-hearted there is a *C. fusca* var. *violacea*, which is hairless and a very pretty deep violet but seems to me to miss the point rather.

There are a number of other specific epithets for brown, not all of them valid. It is possible that *hepaticus* was once used to mean liver-brown (from the Greek for liver, *hepar*), but the suggestion in an RHS publication that this is exemplified by *Anemone hepatica* is wrong: the species name refers to the shape of the leaves (see 'Anatomy'). Other browns include *badius*, meaning chestnut brown, a colour that can be seen in the extraordinary clover *Trifolium badium*, the flowers of which open a clear yellow and then change to a dark, shining brown, sometimes appearing two-tone like a cartoon bumblebee. Among the reddish browns are *rufus* (think William II), as in the Himalayan cherry *Prunus rufa*, grown chiefly for its peeling rich-brown bark, and the bamboo *Fargesia rufa*, which has new culms tinged a bright rust. A similar colour is defined as

russatum, from *russus*, another Latin word for red and familiar in English as 'russet'. *Rhododendron russatum* is so called
because it has red-brown scales covering the undersides of
the leaves. Similarly, and fairly obviously, *cinnamomeus* is
cinnamon-coloured, as in the fronds of the fern *Osmunda
cinnamomea*.

Flowers that we describe as 'black' are usually a very dark
purple, and the specific epithets from the Latin word *niger*
rarely refer to them – and indeed often promise more than they
deliver. One exception is the black false hellebore, *Veratrum
nigrum*, the name of which doubles up the darkness, since the
genus name is from the Latin *vere* = truly + *ater* = black. It is
the dark plum flowers growing on tall, dark grey stems that
give the plant its dusky botanical name. Elsewhere *niger* and
nigrus tend to refer to stems, fruit or foliage. Occasionally in
plants there is something truly black, like the shining boot-
polish black of the bamboo *Phyllostachys nigra*, whereas the
stems of *Hydrangea macrophylla* 'Nigra' only *appear* black
and are in fact a very dark purple. The black walnut, *Juglans
nigra*, owes its name to its dark grey bark, which is also deeply
furrowed, unlike that of the common walnut, *J. regia*, which is
smooth. However, from a distance, especially when the tree is
in fresh green leaf, the trunk can look almost black. As already
mentioned, it is the purple-black fruit rather than the leaves or
stems of the elder *Sambucus nigra* that give it its name, and the
same goes for the black mulberry, *Morus nigra*.

Rather more baffling are the names of the black hellebore,
the black allium, and the black mullein. *Helleborus niger* is also
known as the Christmas rose because it traditionally produces
flowers in the depths of winter. These flowers are, however,
the purest white. An explanation of the name that owes rather
more to high literary style than to botanical taxonomy is

provided by Reginald Farrer: *H. niger* is 'so called because its heart, or root, is black, while its face shines with a blazing white innocence unknown among the truly pure of heart'. Farrer does not, alas, provide similar explanations for the naming of *Allium nigrum*, which also has white flowers, or *Verbascum nigrum*, with its tall candles of bright yellow. The botanical and common names of the black allium refer to the prominent ovaries at the centre of each flower, which are in fact a dark green but can look black from a distance. The two names of the black mullein have long puzzled naturalists, so much so that James Sowerby in his 36-volume *English Botany* (1790–1814) exclaimed: 'By what figure of speech this beautiful plant can be called black, not having a particle of colour about it, we will not determine. All the old botanists, however have denominated it: and if they had any meaning, it can only have been that it was not white.' As with the hellebore, the reason for the name is hidden underground and *nigrum* refers to the plant's black roots.

The genus name of the beautiful blue annual sometimes called love-in-a-mist has a similar etymology and is *Nigella* because of its jet-black seeds. Also related to the Latin word for black are *nigrescens* and *nigricans*, which mean becoming or turning black. Once again the yellow-flowered broom *Cytisus nigricans* owes its name to its black roots, whereas the

Love-in-a-mist, *Nigella damascena*

more apparent purple-black shoots and leaves that unfurl a very dark green explain the naming of *Hosta nigrescens*. *Ophiopogon planiscapus* 'Nigrescens' is a plant that is not so much *becoming* black as fully arrived, even in its seedlings. The liquorice straps of its leaves are about as truly black as you get in a garden, a perfect foil for the small lilac flowers, which are followed by clusters of fruit like shiny black aniseed balls.

The Greek word for black is *melas*, giving us the prefix *melano-* used to describe various parts of plants. The African oil palm *Elaeis guineensis* was originally called *E. melanococcus* (meaning black berry) because of its large clusters of black fruit. Meaning much the same, *melanocarpus* (black fruit) is more common, as in the unusually black-fruited *Cotoneaster melanocarpus* or the black chokeberry *Aronia melanocarpa*.

To end with there are some colour epithets that sound vague and cause disagreement about what they actually mean. Among these are *luridus*, which in Latin means 'sickly yellow, sallow, wan, ghastly' but in botany supposedly means anything from 'smoky' or 'dingy brownish yellow' to 'dirty brown, a little clouded'; *tristis*, the Latin word for sad, meaning sombre-coloured; *pullus*, the Latin for dingy or sombre, which ranges botanically from 'dark coloured' and 'dusky' to 'raven-black, almost dead black'; and *phaeus*, from the Greek *phaios*, meaning dusky. The difficulties of *luridus* can be seen in the iris-like flowers of the Cape tulip *Moraea lurida*, which seem to display a wide range of colours, including pale yellow and bright, not in the least dirty, clouded or indeed lurid, brown and maroon. Pretty as they undoubtedly are, the flowers have a peculiarly revolting smell designed to attract the flies that act as the plant's principal pollinators. The notion of a sad plant is not perhaps one many gardeners would wish to entertain, but the drooping stance this evokes is in fact borne out when *tristis* is applied to shrubs and trees with a 'weeping' habit. (See 'Anatomy'.) As a colour, *tristis* is also highly variable, as may be seen by comparing the truly sombre, almost black bells of *Fritillaria affinis* var. *tristis* (sometimes var. *tristulis*) with the distinctly wan greenish-cream flowers of *Gladiolus tristis* or the rather murky ones of *Pelargonium triste*, which are a dingy yellow or ivory with markings ranging from pale pink-purple to dark maroon. As we know, raven-black flowers don't really exist, and the usual example given for *pullus* is *Campanula pulla*, the flowers of which are a beautiful, very dark violet. The *phaeus* epithet is very familiar from the dusky cranesbill, *Geranium phaeum*, sometimes called the mourning widow because of its very dark purple, almost brown flowers. Once something of a rarity, this plant became hugely popular when there was a vogue for 'black' flowers. It is also a plant you need never buy twice since it seeds very freely, though the results can be variable in colour. Assorted cultivars have flowers ranging from mauve through violet to maroon, and there is even a white 'Album' cultivar, which Christopher Lloyd quite rightly dismissed because it has none of the vigour of the species and is inclined to keel over in, rather than illuminate, those dark corners of the garden for which it is rashly recommended. The cultivar 'Samobor' has leaves splotched with maroon, but such is the promiscuity of *G. phaeum* that this foliage variation seems to appear without selection if the mere species is given its head.

It will have become apparent that Latin epithets denoting colour are often as tricky to pin down as English ones. As Reginald Farrer observed: 'Rose-white, pearl-white, ice-white, snow-white, chalk-white, wax-white, are all totally different tones; but as no three people see ice or snow or wax or pearl or rose with the same eyes, or have the same idea of them, the descriptive tag can only be the inventor's own particular little pet view, having no help or authority except for himself.' While most us can agree about primary colours, the huge array of tones and shades we encounter in a garden makes any attempt to tie down specific epithets to individual gradings of colour a doomed enterprise. It should therefore be admitted that the following lists are at best a rough guide.

Word Lists

RED AND PINK

atro-fuscus = dark red-brown

atrorubens = dark red

atrosanguineus = dark blood red

cardinalis = scarlet

carmineus = carmine

carneus = flesh-coloured, usually pink

cerasinus = cherry red

cinnabarinus = cinnabar

cinnamomeus = cinnamon brown

coccineus = scarlet

corallinus = coral red

cruentus = blood red

cupreus = coppery red

erubescens = reddening, blushing

erythr- (prefix) = red

flammeus, flammeolus = flame-coloured

fulgens, fulgidus = shining, glistening, and usually red

fuscoruber = dark red

igneus = flame red

incarnatus = flesh-coloured

kermesinus = (unreliably) carmine

lateritius = brick red

miniatus = vermilion

phlogiflorus = with flame-coloured flowers

phoeniceus = pure red

puniceus = pure red

rhodo- (prefix) = light red

rhodanthus = with rose-coloured flowers, red or pink

roseus = rose pink

rubellus = light red

ruber, rubescens, rubeus, rubicundus = generic red

rubiginosus = rusty red

rubromarginatus = with red margins

rutilans, rutilus = metallic red

sanguineus = blood red

scarlatinus = scarlet

vermiculatus = vermilion

ORANGE

aurantius, aurantiacus = pure orange

cupreus, cupreatus = copper-coloured

fulvus, fulvescens, fulvidus = tawny or burnt orange

testaceus = terracotta

vulparius = fox red

YELLOW AND GOLD

armeniaceus = apricot

auratus, aureatus = decorated with gold

aureus = golden

aureo- (prefix) = golden

aureomaculatus = spotted gold

caloxanthus = beautiful yellow

canarinus = canary yellow

cereus = corn yellow

cerinus = wax yellow

chrys-, chryso- (prefix) = yellow or golden

chrysanthus = yellow-flowered

chrysographes = marked with gold

citrinus, citreus = lemon yellow

croceus = saffron yellow

flavescens, flavicans, flavidus = yellowish

flavens = yellow

flavissimus = very yellow

flavovirens = yellowish green

flavus = sulphur yellow or pure yellow

icterinus = jaundice yellow

isabellinus = dirty yellow

luteolus, lutescens = pale yellow

luteus = generic yellow

ochraceus = brownish yellow

squalens, squalidus = dull or dirty yellow

stramineus = straw-coloured

succineus = amber

sulphureus = sulphur yellow

vitellinus = egg-yolk yellow

xanth-, xantho- (prefix) = yellow

xanthinus = yellow

GREEN

aeruginosus = verdigris-coloured

atrovirens, atroviridis = very dark green

chlor-, chloro- (prefix) = green

chlorinus = yellow-green

chloroticus = pale green

flavovirens = yellowish green

glaucus, glaucescens = blue-green

melanochlorus = very dark green

nigro-virens = green-black

olivaceus = olive green

prasinus = leek green

psittaceus = parrot green

saligineus = willow green

smaragdinus = emerald green

virens, viridis, virescens, viridescens = degrees of clear green

viridiflorus = green-flowered

viridissimus = very green

viridulus = rather green

viridistriatus = with green stripes

xanthochlorus = yellowish green

BLUE

azureus = azure

caelestis = sky blue

caeruleus = bright cornflower blue

caerulescens = bluish

caesiglaucus = blue and glaucous

caesius = greyish blue

cobaltinus = cobalt blue

coelestinus, coelestis = sky blue

cyaneus, cyanus = clear bright blue, cornflower blue

cyanthus = blue

lazulinus = ultramarine

INDIGO, VIOLET AND PURPLE

amethystinus = deep violet

atropurpureus = very dark purple verging on black

atroviolaceus = dark violet

hyacinthinus = deep violet

ianthinus = violet blue

lilacinus = lilac

malvinus = mauve

porphyreus = porphyry

purpureus = purple

purpurascens, purpurellus, purpurinus = purplish

tyrianthinus = purple-violet

tyrius, ostrinus, blatteus = royal purple

vinaceus, vinosus, vinicolor = wine-coloured

violaceus, violascens, violeus = violet

White

albescens = whitish, becoming white

albomaculatus = white-spotted

albovariegatus = variegated white

albus, albidus = white

candens, candicans, candidus = shining white or pure white

candidissimus = very white

canescens = with off-white hairs

canus = off-white

dealbatus = dusted with white

eburneus = ivory white

lactiflorus = with milk-white flowers

leucanthus = white-flowered

leuconeurus = white-veined

leucophaeus = light grey

nivalis, niveus = snow white

Silver and Grey

argentatus = silvered

argenteo-guttatus = silver-spotted

argenteo-marginatus = silver-margined

argenteovariegatus = variegated silver

argenteus = lustrous silver

argyrophyllus = silver-leafed

argyrotrichon = silver-haired

canescens = hoary or greyish white

cinereus, cinerarius, cinerascens = ash grey

ferreus = iron grey

fumeus, fumidus = smoky

griseus = pure light grey
incanus = quite grey, hoary
leucophaeus = light grey

murinus = mouse grey
plumbeus = blue-grey
schistaceus = slate grey

BROWN

brunneus = solid brown
brunnefolius = brown-leafed
castaneus = chestnut brown
cinnamomeus = cinnamon
ferrugineus, ferruginosus = rust
   brown
fulgineus, fulginosus = sooty,
   dark brown
fuscus = swarthy

glandulaceus = tawny
   yellow-brown
luridus = dirty, clouded brown
   or yellow
phaeus = dusky
rufus, rufescens = rufous,
   reddish brown
russus = brownish black
umbrinus = umber

BLACK

anthracinus = coal black
ater = dead black
aterrimus = deep black
atro- (prefix) = dark
atratus = blackened
carbonaceus = coal black
coracinus, corvinus = crow
   black, shiny black
denigratus = blackened
ebenaceus, ebenus = ebony

   black
mela-, melano- (prefix) = pure
   black
nigellus = black
niger, nigrus = black
nigrescens, nigricans =
   blackish, becoming black
nigritus = dressed in black
piceus = brownish black
pullus = dusky

ALSO

ardens = glowing

bicolor = two-coloured

obscurus = indistinct, dark

mutabilis = changeable of
colour

nitens, nitidus = shining

pallidus, pallidiflorus = pale,
pale-flowered

tinctorum, tinctorius = dyers',
used by dyers

tinctus, tingens = coloured

tricolor = three-coloured

tristis = sombre-coloured

versicolor = various colours

variegatus = variegated

ATLAS

One of the ways in which early botanists distinguished between different species of plants was to give them names relating to where they grew in the wild. Plants could be identified by general habitat – species that grew in woodland or those that grew on mountainsides – or by a particular country or region. Some of the specific epithets for habitat, such as *sylvestris* and *alpinus*, will be very familiar to gardeners because widely used; others, such as *ammophilus* or *clivorus*, less so. Botanists have created words for almost any habitat in which you can imagine a plant growing, but we will concentrate on the ones you are most likely to come across when sauntering round a nursery.

Some of these are so vague as to be more or less useless to gardeners, though not to taxonomists. *Borealis*, for instance, means northern, taken from *boreas*, which is what Greeks called the north wind – as in the Aurora Borealis or Northern Lights. The Latin word for southern is *austrinus*, and the botanical epithet means the same – but can be more exact, as we'll discover when we discuss Australia. Thus, the alpine flea-bane, which is native to Greenland and parts of north-eastern

Canada, is *Erigeron borealis,* while the bastard white oak, which is native to the southern states of the USA, is *Quercus austrina.* The other two points of the botanical compass are *occidentalis* (west) and *orientalis* (east). The western epithet often means America or the American West, so that the American syca-more is *Platanus occidentalis,* and the western or Californian redbud is *Cercis occidentalis,* distinguishing it from the closely related *C. canadensis,* which grows in north-eastern America.

The eastern epithet is far more widely used both because 'the Orient' had a long historical grip on the European imagination and because a great many plants came from this fabled place. Just as 'the Orient' was a blanket term covering both the Near and Far East, so *orientalis* has a wide geographical application. *Helleborus orientalis,* for example, is a native of Greece and Turkey, and while the former country seemed, at any rate to the British imagination, 'western', the cradle of Classicism and democracy, the latter was for many centuries regarded as distinctly, if alluringly, 'eastern', embodying lassitude and des-potism. The French agreed: 'What can a Man say of a Country inhabited by *Turks*?' Joseph Pitton de Tournefort asked rhet-orically when setting out on his journey to the Levant in March 1700. 'Almost their whole Life is spent in Idleness: to eat Rice, drink Water, smoke Tobacco, sip Coffee, is the Life of a *Mussulman.*' Tournefort, who was professor of botany at the Jardin du Roi in Paris, had been sent into the Ottoman Empire by Louis XIV. The aim of the journey was to make 'pertinent Observations, not only upon the natural History, and the old and new Geography of those Parts, but likewise in relation to the Commerce, Religion, and manners of the different People inhabiting there', all of it recorded in Tournefort's hugely enjoyable *Relation d'un voyage du Levant* (1717), translated into English the following year as *A Voyage into the Levant.*

Tournefort was the first person to record *Rhododendron ponticum* and the common azalea, *Rhododendron luteum*, both encountered at Trebizond. At Erzerum in Armenia he discovered the oriental poppy, *Papaver orientalis*, reporting that 'by way of delicacy' the indigenous population 'eat the Heads of it when they are green, tho very acrid, and of a hot Taste'. He sent seeds of the plant back to Paris, and from there some were sent on to England to be raised by George London at the Brompton Park Nursery in the early years of the eighteenth century. Tournefort gave the species name *orientalis* to a large number of other plants discovered on his travels, including an echium, an artemisia and a thistle. Some of these names, such as *Ranunculus orientalis*, are still in use, but a great many more have been discarded. Even so, the hyacinth, the Turkish iris, a clematis and leopard's bane are among the plants that take the epithets *orientalis* or (for neuter genus names) *orientale*.

There are other epithets that sound more geographically exact than they are. *Alpinus*, for example, does not necessarily denote plants growing in the Alps. It's the same when we talk about 'alpine plants' or 'alpines': *alpinus* simply means a plant that grows in mountainous regions, technically above the timber line. Compare this with *alpestris*, which refers to the lower slopes of mountains, below the timber line. A rather less precise epithet is *montanus*, meaning 'of mountains', most familiar from *Clematis montana*, which grows happily – not to say uncontrollably – in British gardens but is native to mountainous regions in Asia. A similarly general, though far less common epithet is *acraeus*, meaning growing at high altitudes, as in the pink-flowered South African *Pelargonium acraeum*, or *Ranunculus acraeus*, which is known as the alpine buttercup but is in fact native to a mountainous region of New Zealand.

Many of the plants growing at such heights thrive in crevices of rock, from which we get *rupestris* and *saxatilis*, from *rupes* and *saxum*, two Latin words for rock or boulder. Some dictionaries suggest that *rupestris* means growing *on* rocks, whereas *saxatilis* means growing *among* them, but essentially they mean the same thing, indicating plants particularly suited to growing in rockeries. *Erodium rupestre*, for example, is native to the limestone regions of north-eastern Spain, and golden alyssum, *Aurinia saxatilis*, is native to rocky areas in central Europe. A related epithet is *rupifragus*, meaning to break rocks, as in the pale orange poppy *Papaver rupifragum*, which despite its delicate appearance can indeed even push its way up through garden paving. The genus *Saxifraga* is similarly named because in the wild saxifrages grow in crevices, giving the appearance of having burst out of solid rock.

Coming a little further down the slopes we arrive at *collinus*, from the Latin *collis*, a hill, and *clivorus*, from the Latin *clivus*, a slope. What constitutes a hill or slope rather depends upon which part of the world you are in. The hill geranium, *Geranium collinum*, for example, grows in meadows at an elevation of some 3,600 metres in its natural Asian habitat; to put this in perspective, England's highest *mountain*, Scafell Pike, reaches a mere 978 metres. The geranium nevertheless grows happily enough much nearer sea level.

Gardens that are actually beside the sea are often said to be very difficult to cultivate because of the sand or rock, the wind, and the salt air. That said, one of the most famous gardens in Britain is the one Derek Jarman created in the shingle at Dungeness, and I was once shown round a garden on the southernmost tip of Vancouver Island filled with perfectly ordinary plants that had been tucked into pockets of wave-lashed rock projecting right into the sea. For plants that relish rather than merely endure such conditions, the epithet is *maritimus*, from the Latin *mare*, the sea. The pink tufts of thrift, *Armeria marit-ima*, are a familiar sight in the coastal regions of Britain, while one of the most striking plants in Jarman's garden is the beauti-ful dark-purple and cupreous-green seakale, *Crambe maritima*. The aforementioned *ammophilus* denotes a sand-loving plant, from the Greek *ammos* = sand + *philia* = affection; a synonym is *arenarius*, from the Latin *arena*, meaning sand (and conse-quently the sand-strewn area where gladiatorial contests took place). The small evening primrose *Oenothera ammophila* and the shaggy white sand pink *Dianthus arenarius* grow on dunes in the wild and therefore tolerate drought. As you might expect, the epithet *aridus* is used for other plants that like dry places, such as *Penstemon aridus*, a small blue-flowered native of Wyoming that grows best in gardens in stone troughs, and the cream-flowered *Pelargonium aridum*, which survives the summer drought in its native South Africa by storing water in its tuberous roots.

Specific epithets for plants that prefer damp conditions run from *rivalis* and *rivularis* (from the Latin *rivus*, river, and mean-ing growing beside streams), through *palustris* and *uliginosus* (from *palus* and *uligo*, marsh or waterlogged ground) to *amphi-bius* and *aquaticus* (self-explanatory and very wet indeed). The water avens, *Geum rivale*, is a British native with bowed flower-heads now widely grown in bog-gardens, especially in the coppery 'Leonard's Variety' or pale yellow 'Lemon Drops' cultivars, while the equally popular crimson-flowered thistle *Cirsium rivulare*, although not needing its feet kept sopping wet, also prefers moist conditions. Another specific epithet for plants that relish damp places is *bufonius*, from *bufo*, the Latin word for a toad, and *Juncus bufonius* has the common name of toad rush. The *Ranunculus* genus of buttercups is named for similar damp-loving reasons, this word being a diminutive

Marsh marigold or kingcup,
Caltha palustris

of *rana*, the Latin for a frog. Plants liking even boggier plots include the marsh marigold or kingcup, *Caltha palustris*, and the elegant, sky-blue bog sage, *Salvia uliginosa*, discovered in the swamps of South America in the nineteenth century and so given this name, though it will survive in soil that is merely damp. As its botanical name indicates, *Persicaria amphibia* not only grows wild in water meadows and other marshy areas, but also in deep ponds, its leaves and the pink poker flowers characteristic of bistorts floating on the water. Actually requiring a depth of water in which to grow is water mint, *Mentha aquatica*, the strong peppermint scent of which combines evocatively with English mud to provide what Rupert Brooke called 'the thrilling-sweet and rotten / Unforgettable, unforgotten / River-smell' characteristic of his garden in Grantchester.

Some plants thrive in the openness of grassland, while others prefer the dappled shade of woods. For the former there are several epithets, including *campestris* (from the Latin *campus*, a plain or open field), *arvensis* (from the Latin *arvum*, an arable or cultivated field), *agrarius* and *agrestis* (Latin words meaning agrarian or rustic, derived from *ager*, a field), and *pratensis* (from the Latin *pratum*, a meadow). Of these, both *agrarius* and *agrestis* tend to be used of plants such

as the field fumitory, *Fumaria agraria*, and field speedwell, *Veronica agrestis*, that grow wild on agricultural land and are not cultivated in gardens. Plants with the species name *campestris* also belong in the countryside but are cultivated there: for example field mustard, *Brassica campestris*, which forms bright yellow sheets in the English landscape, and the field maple, *Acer campestre*, which is really a hedgerow tree but is also grown in gardens for its autumn colour. A similar migration to the garden has occurred among other plants that grow naturally on arable land, including the field rose, *Rosa arvensis*, a rambler with single pale cream flowers and a central mass of bright gold anthers, and the field scabious, *Knautia arvensis*, which has lilac pincushion flowers worthy of any border. On my family's farm we had an old meadow that had never been cultivated but only used for grazing. As an experiment we kept sheep, cattle and horses off it in the growing season and it very soon became an astonishing *mille-fleurs* tapestry of the sort people strive for in 'wild' gardens. The fact that many plants like growing amongst grasses in meadows explains why the epithet *pratensis* is so common. The meadow cranesbill, *Geranium pratense*, has light violet flowers, but there are now many garden cultivars in different shades of blue, mauve and pink, including the violet-grey 'Mrs Kendall Clark', the double mauve-pink 'Summer Skies', and such variably striped ones as the blue and white 'Striatum' and the bright-purple and white 'Splish Splash'. The yellow-flowered meadow vetchling, *Lathyrus pratensis*, has yet to leap into gardens, but lady's-smock or cuckoo-flower, *Cardamine pratensis*, certainly has – though some would argue that it still looks at its best when its washed-out-pink flowers are stippling a meadow of lush green grass.

Wilder than even uncultivated fields are heaths, for which

the Greek word is *ereike*, from which we get the heather genus *Erica*. Another genus that may have the same etymology is *Hypericum*, but it has also been suggested that its name derives from the Greek for above (*hyper*) + picture (*eikon*), because the flowers were hung over images in people's houses to keep away the devil at the midsummer festival of St John's Eve, when fires were lit for a similar purpose. It is for this connection that the plant was known by 'barbarus writers' in Turner's time as Fuga demona and is still called St John's wort – but as justification for supporters of the first derivation some species of the plant do grow on heathland.

Perhaps even more woodland natives have been cultivated in British gardens than field species. And when they are as pretty as the wood anemone, *A. nemorosa*, with its pale flowers flicking the dark backs of their petals in a breeze, this is hardly surprising. The epithet *nemorosus* is a Latin word meaning well-wooded, and is also the species name of *Salvia nemorosa*, grown in gardens in a large number of cultivars. The two other epithets for woodland plants are *sylvaticus* and *sylvestris*, from *silva*, a Latin word for a wood or forest. Another woodland sage, *Salvia × sylvestris*, is also found in gardens in a large number of cultivars, including ones with pink rather than purple or blue flowers. The common beech is *Fagus sylvatica* because it is a woodland tree. The plant that, despite its species name, is perhaps most hard to imagine growing in a wood or forest is the tall and huge-leaved *Nicotiana sylvestris*, with its long, tubular and highly fragrant white flowers, introduced from the Andes in the early 1800s. A great many other plants native to woodland are grown as cultivars, notably the woodland cranesbill, *Geranium sylvaticum*, and cow parsley, *Anthriscus sylvestris*, which is now available with dark purple ('Ravenswing') or gold ('Golden Fleece') foliage – though the so-called pink

cow parsley, which has lovely flat heads of rose flowers, is in fact the cultivar of a species of the unrelated hairy chervil, *Chaerophyllum hirsutum*.

We now move on to those plants that botanists sorted by country of origin. A catch-all epithet is *barbarus*, the Latin word for 'foreign', and not to be confused with *barbatus*, which means bearded. It survives for the Chinese box thorn, *Lycium barbarum*, which despite its English name has a wide geographical range from China all the way west to Europe. Also known, after the aristocrat who introduced it to Britain in 1730, as the Duke of Argyll's tea tree, this scrambling shrub has attractive light-purple flowers but is better known for its bright orange-red fruit, widely sold in dried form as goji berries.

Clearly an epithet designating merely 'foreign' origins is of little use (and indeed is not much used), whereas ones relating to particular countries or regions are more helpful in differentiating between species. Different kinds of rock rose, for example, might be sorted in this way: *Cistus creticus* from Crete, *C. libanotis* from Lebanon, *C. monspeliensis* from Montpellier, and so on. This seems simple enough, but is not entirely defining for those many plants that grow wild in several different parts of the world. Indeed, *Cistus monspeliensis* grows in other parts of southern Europe and in North Africa too, and a kind of *Cistus creticus* grows in Corsica. The solution in the latter case was to create a subspecies, *C. creticus* subsp. *corsicus*. A further problem is that over the centuries countries have come and gone as empires rose and fell, and regions have changed their names. To take a plant everyone knows even if they don't grow it: the species name of *Rhododendron ponticum* indicates that it was native to Pontus in Asia Minor, but few gardeners who are not also historians of the ancient world would know where that was. Pontus is no longer on the map any more than Asia Minor is; it was in fact an area on the southern edge of the Black Sea in what is now the Anatolian region of present-day Turkey.

It is something of a relief that botanists do not revise names when political and geographical boundaries change, but this does mean that a sizeable proportion of species names refer to countries and regions that have been lost in time. The specific epithet *baeticus*, for example, is used for plants native to Andalusia, but you would hardly know this unless you also knew that Baetica was the name given by the Romans to an imperial province in what they called Hispania Ulterior, roughly corresponding to Andalusia in present-day Spain. *Thymus baeticus* helps a little here, since its common name is Spanish lemon thyme; *Cytisus baeticus*, on the other hand, has the geographically imprecise common name of south-western broom, and is indeed also native to southern Portugal, Algeria and Morocco.

In addition, the Latinate versions of some of those place names that do still exist are not always easy to recognise. While most people would be able to deduce that *Salvia turkestanica* comes from Turkestan or that *Genista hispanica* originates in Spain, epithets such as *suecicus* (from Sweden) or *siculus* (Sicilian) are perhaps less easy to guess. Other geographical names can be all too easy to confuse. Unless you are an alumnus of Cambridge University, for example, you might well mix up *cantabricus* with *cantabrigiensis*. The white winter-flowering alpine daffodil *Narcissus cantabricus* is native to Cantabria in northern Spain; the pale yellow single-flowered *Rosa × cantabrigiensis* is a natural hybrid between *R. sericea* and *R. hugonis* which was isolated and bred at the Cambridge Botanic Garden. What is more, some names are not what they might appear.

The specific epithet of the cedar *Cedrus atlantica* does not refer to the ocean or the Atlantic seaboard of America but to the Atlas Mountains. And the popular Michaelmas daisy that we are now supposed to call *Symphyotrichum novi-belgii* does not come from Belgium but from New York State, named Nova Belgica by Dutch settlers in the seventeenth century after the ancient Belgic tribes of the Low Countries.

If you haven't by now already thrown your map of the world onto the compost heap in despair, you may well do when you learn that both the common and botanical names of plants sometimes came about by mistake and are geographically 'wrong'. The most egregious example of this is the Japanese anemone, *Anemone japonica*. The plant was found growing in Japan in the 1680s by a German doctor called Andreas Cleyer, who was sent by the Dutch East India Company to their trading post of Dejima, a tiny artificial island that had been created in the Bay of Nagasaki in order to corral foreign traders and prevent them from spreading their religion amongst the indigenous population. Cleyer nevertheless managed to explore the local flora and fauna, among which he found and recorded a tall kind of anemone that was growing in the wild. It was Linnaeus's pupil Carl Peter Thunberg who first named the plant a century later in his *Flora Japonica* (1784), and it would be another sixty years before Robert Fortune, a Scottish botanist working for the British East India Company, found the same plant growing wild in China. It transpired that the species had been introduced into cultivation in Japan many years before Cleyer discovered it there and that the 'wild' specimens he saw were in fact garden escapes. In 1908 the plant was finally given the more correct name of *Anemone hupehensis* after it had been found growing in the central Chinese province of Hupeh (now Hubei). By this time, however, breeders had crossed the plant with *Anemone vitifolia*, a native of both India and China, which Lady Amherst, wife of the Governor General of Bengal, had introduced from Nepal in around 1829, when she also introduced the ornamental pheasant that bears her name. The result was *Anemone* × *hybrida*, best known in the cultivars 'Elegans', with semi-double pink-mauve flowers, and 'Honorine Jobert', with single white flowers and named after the daughter of the French horticulturist who propagated it in the early 1850s. All these plants – species, hybrids and the geographically bet-hedging *Anemone hupehensis* var. *japonica* – are still referred to as Japanese anemones, though feeble and thankfully futile attempts have been made to rebrand them as 'Chinese anemones'.

Other plants designated *japonicus* that were first found in China include the yellow-flowered shrub *Kerria japonica*, which is native to both countries, as well as to Korea. It was named after William Kerr, who had been selected by Joseph Banks in 1803 to travel to China with David Lance, a superintendent of the East India Company's trading post in Canton. Lance had offered to find plants for Kew, but he needed a horticulturally trained assistant. Kerr found and dispatched a large number of plants to England, including the tiger lily, *Lilium tigrinum*, the shrubs *Pieris japonica* and *Pittosporum tobira*, and the pale yellow climbing rose that would be named after his employer's wife, *Rosa banksiae*. Since Kerr was a lowly gardener, being paid no more than his usual annual salary of £100, these introductions were credited to Lance or to the East India Company, but amends were made in 1816, two years after Kerr's death, when *Corchorus japonica* was renamed *Kerria japonica*. In fact Kerr had been sufficiently well regarded by Banks to be appointed curator of the botanic garden in Colombo, in what was then Ceylon, a post in which he served for only two years, his death perhaps a result of the opium he was reported to have taken to smoking.

A plant that has confusingly taken 'japonica' as one of its common names is the flowering japonica, also known as flowering quince. It was first named *Pyrus japonica* (thus placing it in the pear genus) in 1784 by Thunberg, who had found it growing in the mountainous Hakone region of Japan. A few years later Banks introduced a plant from China which he thought the same as Thunberg's Japanese one, a mistake supposedly rectified in 1818 when Banks's plant was renamed *Pyrus speciosa*. Unfortunately the Chinese plant was still being referred to as *P. japonica* fifty-one years later when the Bristol nursery of W. Maule and Son introduced Thunberg's original Japanese one to Britain. In order to avoid confusion, but in fact merely compounding it, this plant was renamed *Pyrus maulei*. Meanwhile, a French botanist had decided that these plants were not pears but quinces, which the fruits more closely resemble in shape and scent, so they became *Cydonia*; but eventually they were placed in the *Chaenomeles* genus, Thunberg's original plant becoming *Chaenomeles japonica* and Banks's interloper *C. speciosa*. The final error in this saga of muddled naming is that *Chaenomeles* is from the Greek *khaino* = to gape + *melos* = apple, because it was incorrectly reported that the fruit (not, as we have seen, an apple) splits – and therefore 'gapes' – when ripe. It is perhaps hardly surprising that gardeners throw their hands in the air and simply call both plants 'japonica'.

It is, however, in Japan that we are going to start a brief turn around the world and its botanical epithets. In addition to *japonicus*, plants of Japanese origin sometimes take the species name *nipponicus*, from Nippon, the Japanese name for the country which more or less translates as 'land of the rising sun'. Given the Japanese passion for flowering cherries, it seems invidious for a single species to be named *Prunus nipponica*, even though it is a particularly beautiful one with clear light-pink single flowers, and foliage that bursts into flaming yellow and orange in the autumn. Although the chrysanthemum is the floral emblem of Japan, stylised as the seal of the imperial court and widely used in the decorative arts, one of its least distinguished species has been honoured with a name meaning 'the Japanese flower of Japan'. Originally *Chrysanthemum nipponicum*, this shrubby daisy is now *Nipponanthemum nipponicum*. It has glossy green leaves and yellow-centred white flowers that bloom in late autumn but are not otherwise particularly alluring. In addition to *japonicus* and *nipponicus* we have *yedoensis* and *yezoensis*, both of which are derived from Yedo (sometimes transliterated as Edo), a former name of Tokyo; but at best these mean nothing more specific than native to Japan. The hybrid Yoshino cherry *Prunus* × *yedoensis* 'Somei-Yoshino', for example, has no particular connection to Tokyo and takes its cultivar name from a district of Japan which is some 600 kilometres south-west of the capital. The clear yellow but frost-tender day lily *Hemerocallis yezoensis* is native to the Kuril Islands, the sovereignty of which is disputed between Russia and Japan, while *Viola yezoensis* is a pretty little violet with dark bronze foliage and purple-streaked white flowers sometimes called the Chinese violet because it is also native to China.

The most frequently used epithet for China is *sinensis*, from the Greek word for the Chinese people, *sinai*, but an anglicised alternative is *chinensis*. The origins of the tea plant, which the aforementioned Robert Fortune illegally exported from China to Darjeeling on behalf of the East India Company, thus inaugurating the huge Indian tea trade, are enshrined in its botanical name, *Camellia sinensis*. Other familiar plants from China include the orange tree (*Citrus sinensis*), wisteria (*Wisteria sinensis*) and the Chinese winter hazel (*Corylopsis sinensis*) with its wonderfully fragrant cowslip-yellow racemes of flowers. The regular Chinese hazel takes the alternative epithet and is *Corylus chinensis*, as does the shrub *Enkianthus chinensis*, which has delicate clusters of dangling bell-shaped flowers of a creamy yellow streaked with pale red. Alongside *hupehensis* are specific epithets pertaining to various regions or provinces of China. The historical north-eastern region of Manchuria gives us the white birch *Betula mandshurica*, grown chiefly in Britain in the Japanese variety, var. *japonica*. Other plants from the region include the aniseed-scented white-flowered herbaceous *Clematis mandshurica*; the Manchurian maple, *Acer mandshuricum*; and the Manchurian apricot, *Prunus mandshurica*. An alternative epithet is *manchuriensis*, used for an alpine saxifrage and the catmint *Nepeta manchuriensis*, of which the tall violet 'Manchu Blue' cultivar is a striking alternative to the more ubiquitous *Nepeta* 'Six Hills Giant'.

The south-western Yunnan Province was a particularly popular hunting ground for western plant collectors, including several of the French missionaries who did so much to bring Chinese plants to Europe, and numerous plants bear the epithet *yunnanensis*. The Yunnan lilac, *Syringa yunnanensis*, was discovered by Père David in 1891 and introduced to Britain around fifteen years later by George Forrest (1873–1932), who collected plants for A. K. Bulley, a Liverpool businessman who founded both the Ness Botanic Gardens (1898) and Bees Nursery (1903) in Cheshire. The winter-flowering *Osmanthus yunnanensis* was also introduced by Forrest in the 1920s, while the pink-flowered Yunnan rhododendron, *R. yunnanense*, was discovered by Père Delavay and introduced to France in 1889. Among herbaceous plants are the white-flowered Yunnan meadow rue, *Thalictrum yunnanense,* the tall cream-flowered Yunnan loosestrife, *Lysimachia yunnanensis,* and the giant Yunnan lily, *Cardiocrinum giganteum* var. *yunnanense,* the flowers of which are streaked with brown-purple and more open than the species. The epithet *formosanus* (not to be confused with *formosus,* which simply means beautiful) is taken from Formosa, the old name for the island of Taiwan, to which the People's Republic of China currently lays much-resisted claim. Exotic plants native to Taiwan include the large white-trumpeted Formosan lily, *Lilium formosanum,* the spotted toad-lily, *Tricyrtis formosana,* and the spectacular carpet-forming peacock orchid, *Pleione formosana,* with bright lilac petals framing a fringed, brown-and-yellow-speckled tubular mouth.

A number of very fine plants are native to Korea, for which there are three specific epithets, two of which are used for different species of geranium. *Geranium koreanum* has large lilac-pink flowers with white centres and deeply toothed and attractively marbled leaves that turn orange-red in the autumn. *G. koraiense* also has leaves with a good autumn colour, and bright pink flowers with a dark and very distinct, almost architectural tracery of veins. *Clematis koreana,* though a more recent introduction, is similar in habit and flower shape to *C. alpina* and is the parent of several widely grown cultivars, of which the pale primrose-coloured 'Amber' was declared Plant of the Year at the Chelsea Flower Show in 2016. The Korean day lily, which has trumpets the colour of a proper farmyard egg yolk, takes the country's third epithet and is *Hemerocallis coreana.*

Whereas Japan and China were at times closed to foreigners, making it difficult for plant hunters to go about their business,

Britain's long if troubled association with India gave travellers a virtual *passepartout* to explore the subcontinent. As early as the 1690s the East India Company surgeon Samuel Browne was sending back seed to both the Oxford Botanic Garden and the Chelsea Physic Garden. The Company's own Botanic Garden was founded at Calcutta in 1786, and its second director was William Roxburgh (1751–1815), one of the first of a distinguished line of Scottish botanists who came to India and sent plants back to Britain. Roxburgh also employed local artists to make botanical drawings, and by the time he retired in 1813 his catalogue of plants growing in the Calcutta Botanic Garden listed 3,240 species. The catalogue was titled *Hortus Benghalensis*, and the latter word became the specific epithet for plants native to Bengal, such as the banyan tree, *Ficus benghalensis*, a vast specimen of which was and remains a famous feature of the Garden.

The specific epithet for India more generally is *indicus*, perhaps familiar to some readers from *Cannabis indica*, otherwise marijuana. Similarly named plants that are actually legal to grow include the blue morning glory *Ipomoea indica*, the Indian horse chestnut, *Aesculus indica*, and *Chrysanthemum indicum*, a rather dreary species in the wild which has nevertheless been hybridised for gardens to produce flowers with a much wider and brighter range of colours than the original yellow. Among regions of the subcontinent, *sikkimensis* fairly plainly denotes plants from Sikkim in the eastern Himalayas, and is used for species such as the Sikkim spurge, *Euphorbia sikkimensis*, and the Sikkim cowslip, *Primula sikkimensis*. Plants from the Himalayan region more generally are indicated by the epithets *himalayensis* and *himalaicus*, and a good deal less obviously by *emodi* and *emodensis*, which refer more particularly to the western part of the range, known in classical times as the Emodus

Mountains. Apart from the blue Himalayan poppy, *Meconopsis betonicifolia*, which makes no reference to its origins in its botanical name, the most familiar of the Himalayan plants is *Geranium himalayense*, grown in British gardens in several cultivars. The very tall and pure white foxtail lily *Eremurus himalaicus* makes a spectacular vertical marker in a border. Also white-flowered and one of the taller members of its genus is the Himalayan peony, *Paeonia emodi*, while the decorative grass *Calamagrostis emodensis* is distinguished from others of its genus by the weeping habit of its plume-like flower-heads.

Kashmir also has several different epithets, most recognisable in a name such as *Iris kashmiriana*, a white-flowered species that is sometimes grown in burial grounds in its native country, where it is called *mazamond*, from *mazar*, a Kashmiri word for a graveyard. The epithet is sometimes rendered *cashmirianus*, as in the white-berried Kashmir rowan, *Sorbus cashmiriana* (sometimes spelled *cachmiriana*), which has very beautiful pale pink single flowers. *Phlomis cashmeriana* has a different spelling again and is an attractive herbaceous species of a genus that can be very coarse; it has spires of lavender-coloured flowers rather than the rather murky yellow ones of the shrubby Jersualem sage, *P. fruticosa*. (The *Phlomis* genus, incidentally, derives its name from *phlox*, the Greek word for flame, because its woolly leaves were used in the ancient world as lamp-wicks.) The Kashmiri variety of the drumstick primula is *Primula denticulata* var. *cachemeriana*, while the Kashmir gentian, which has intensely blue flowers and grows in rocky crevices in the wild, takes yet another specific epithet and is *Gentiana cachemirica*.

Wedged between India and China is the Himalayan republic of Nepal, the plants of which are botanically designated *nepalensis* or *napaulensis*. *Lilium nepalense* is a striking species, the trumpets of which have lightly recurved tepals and the unusual colouring of pale green on the outside and russet on the inside. *Potentilla nepalensis* is a popular garden plant, available in a number of cultivars, notably 'Miss Willmott' (see 'Eponyms'), the pale cherry pink flowers of which have dark red centres. The Himalayan poppy *Meconopsis napaulensis* (also known as the satin poppy) was first recorded in 1821 by Nathaniel Wallich, the Danish plant collector who would serve as Director of the Calcutta Botanic Garden for thirty years, creating there what the much-travelled Bishop Heber described in 1823 as 'not only a curious but a picturesque and most beautiful scene [that] more perfectly answers Milton's idea of Paradise, except that it is on a dead flat instead of a hill, than anything which I ever saw'. The satin poppy species has recently been split up, with *M. napaulensis* retained for the plant that has yellow (rather than the more usual blue, purple or red) flowers and is restricted in the wild to central Nepal.

Tibet was another country very popular with European plant hunters in the nineteenth century, and botanists have once again played a set of variations upon its name. *Meconopsis tibetica* is an unusually coloured species of the Himalayan poppy with brilliant maroon flowers, only recently discovered but already offered for sale by specialist nurseries. The Tibetan or giant cowslip (particularly attractive in its coppery red form) is *Primula florindae*, named not after its country of origin but after Florinda, the first wife of the man who discovered it, the twentieth-century plant hunter Frank Kingdon-Ward; *Primula tibetica*, however, is an alpine species of primrose with yellow-eyed pink-purple flowers. Other plants with names acknowledging their Tibetan origin have the slight variant *thibeticus*, as in the Tibetan whitebeam, *Sorbus thibetica*, first described by Père Soulié in the 1890s but not introduced into cultivation until the 1950s, from seeds collected by Kingdon-Ward. The most familiar garden plant from Tibet is *Clematis tibetana*, one of several species with yellow lantern flowers, and easily confused with *C. tangutica*, which originates in China, its species name taken from the Tangut people of Gansu in the north-west of the country, or with *C. orientalis*, which has a large geographical range and was introduced into cultivation in 1731. *C. tibetana* was first discovered in Tibet over 200 years later, but one of the most commonly grown varieties, *C. tibetana* var. *vernayi*, was collected in Nepal and is listed by the RHS as 'Chinese clematis'. It is of course not unusual for plants found growing in the wild in one country to ignore national boundaries and also thrive in neighbouring states, but this is surely an example of a common name that is less than helpful? *C. tibetana* 'Orange Peel' is also a popular garden plant, its cultivar name referring not to the colour but to the fleshy thickness of the petals (why not 'Lemon Peel'?). Like the Tibetan whitebeam, the pretty pink-flowered Tibetan hellebore adds an 'h' to its species name and is *Helleborus thibetanus*.

Several species names denote places in the Middle East, notably Lebanon, Iran and Syria. The magnificent cedar of Lebanon, which has been one of the most distinctive and elegant trees in British gardens since the seventeenth century, is *Cedrus libani*; a rock rose with bright white flowers is *Cistus libanotis*; and a familiar spring-flowering pink cyclamen is *Cyclamen libanoticum*. Iranian plants take the epithet *persicus*, from the country's former name of Persia, and include another well-known cyclamen, *C. persicum*, with paler flowers than its Lebanese relation and nicely mottled leaves. Another familiar garden plant, though one that infuriatingly never survives in my plot, is *Fritillaria persica*, which has tall steeples of dusky purple-black bells. There is a particularly dismal small-flowered yellow avens called *Geum aleppicum*, which although it has a very wide distribution is named after the Syrian city of Aleppo. The specific epithet for Syria as a whole is *syriacus*: *za'atar*, that mainstay of the fashionable kitchen, is the Arabic word for *Origanum syriacum*, the principal ingredient of what is usually a mixture of several different herbs. The plant itself bears some resemblance to catmint but has pale pink or white flowers. *Hibiscus syriacus* is one of several plants known as the Rose of Sharon, a flower mentioned in the Song of Solomon that has been variously identified as a tulip, a lily and even a crocus. What the Bible plant almost certainly wasn't is the St John's wort often called the Rose of Sharon, *Hypericum calycinum*, which is also known by other biblical names such as Aaron's beard and Jerusalem star. By this time you may not be surprised to learn that *Hibiscus syriacus* in fact comes from India and China; as in the case of *Anemone japonica*, the plant had long been introduced to another country, and in naming it Linnaeus imagined it was a native of Syria, where it had been recorded. In Britain it has the common name of rose mallow, a reference to the pale pink flowers of the species, which are infinitely preferable to some of the 'blue' (grubby violet) and 'chiffon' (messily doubled) cultivars offered by many nurseries.

A very large and handsome red-flowered hibiscus appears in a tub in *Unconscious Rivals*, an 1893 painting by Lawrence Alma-Tadema, who specialised in scenes from classical antiquity and was a very fine painter of flowers. While I am happy to believe that hibiscus was grown in Roman households, I was highly sceptical of the ranks of blue delphiniums in the foreground of Alma-Tadema's *The Finding of Moses* (1904). The infant Moses' basket is hung about with lotuses, which seems right, but could delphiniums really grow in Egypt, or are they merely there to show off the painter's skills? I discovered that delphiniums were indeed sometimes used in the garlands hung about mummies, and although not native to the country may have been imported and grown there. There is definitely, however, an Egyptian yarrow, *Achillea aegyptica*, with bright yellow flowers and finely divided grey-green foliage, and an Egyptian St John's wort, *Hypericum aegypticum*, which in European cultivation is a small, tender alpine, though it can grow much larger in its native country. There is a second specific epithet, taken from the port of Alexandria but also meaning more generally from Egypt: *alexandrinus*, not to be confused with *alexandrae*, which honours Queen Alexandra, the wife of Edward VII. *Senna alexandrina* is a member of the pea family with elegant pinnate leaves and racemes of yellow flowers, and although eminently garden-worthy has long been grown commercially as the laxative commonly known as sennapod (tea made from the leaves has a similar effect). The plant takes its species name from the fact that it was once a major Egyptian export, shipped out across the world from Alexandria.

Other parts of the African continent also yielded plants for western gardens, most notably the Cape, which was often a stopping-off point for European plant hunters en route for the Far East. Among plants with names denoting their South African origin are the Cape figwort, *Phygelius capensis*, a mid-Victorian introduction with orangey-red tubular flowers, and two spectacular evergreen climbers: the flaming red Cape honeysuckle, *Tecoma capensis* (which is not in fact a honeysuckle but has similar flowers), and the stormy blue Cape leadwort, *Plumbago capensis* (syn. *P. auriculata*). Also native to South Africa is the agapanthus, one of the bluest and best of which is *A. africanus*, taking its name from the whole continent rather than a particular region, and introduced to British gardens in the seventeenth century. One would imagine that a plant called *Zantedeschia aethiopica* would hail from Ethiopia, but the arum or calla lily is another native of South Africa. The epithet aethiopicus in fact refers to Africa generally, and *abyssinicus* is used for the country that used to be called Abyssinia, as in the bright blue *Ceratostigma abyssinicum*, which in Britain can only be grown as a conservatory plant.

Algerian ivy, *Hedera algeriensis*, was introduced to Britain in the 1830s and is now chiefly valued in variegated cultivars such as 'Marginomaculata' and 'Gloire de Marengo'. Some nurseries sell a very pale-flowered Algerian subspecies of the Spanish bluebell, *Hyacinthoides hispanica* subsp. *algeriensis*, described as 'exquisite' but in fact merely washy. Altogether more robust and worthy of growing is *Lathyrus tingitanus*, its specific epithet taken from Tingis, a Roman name for Tangier. Commonly known as the Tangier pea, this is a vigorous climber of the sweet-pea genus with bright magenta flowers, although a coral pink cultivar called 'Rosea' is offered to those who share E. A. Bowles's horror of the species' colour. This rather hasty tour of Africa ends offshore on the island of Madagascar, for which the specific epithet is *madagascariensis*. For those who can find room in their gardens for weed-like yellow daisies, the RHS lists *Senecio madagascariensis*, otherwise fireweed or Madagascar ragwort. Altogether more genuinely fiery is the winter-flowering and alas tender *Buddleja madagascariensis*, which at its best has flowers the pure orange of a marmalade cat, beautifully offset by the grey colouring of both its shoots and the undersides of its leaves. The plant was introduced into cultivation as long ago as 1827 and had an RHS Award of Garden Merit somewhat belatedly bestowed upon it in 2002, though the sparsely flowered yellowish specimen photographed for the Society's website hardly recommends it.

The vast area of Russia and Central Asia, across which trade passed along the Northern Silk Road from earliest times, has long been a source of plants for British gardeners. In 1618 John Tradescant the Elder joined an embassy sent by James I to negotiate trade deals with Tsar Michael I of Russia. Neither the embassy nor the plant hunting was very productive, though it may have been from this voyage that Tradescant brought back the Muscovy Rose (*Rosa moscovita*), an unidentified 'purple geranium', and even the larch tree. Muscovy was an old name for the entire Russian Empire rather than Moscow itself, and while *moscovitus* seems to have disappeared from the botanical lexicon (Tradescant's rose was renamed *R. acicularis* because of its sharply pointed leaves), both *russicus* and *ruthenicus* are used for plants native to Russia. The Siberian larch, for example, is *Larix russica* (syn. *L. sibirica*), which is simple enough. More potentially confusing is *Echium russicum*, which the RHS unhelpfully lists as 'red-flowered viper's grass'. This form of viper's bugloss is indeed red-flowered, so people may well confuse *russicum* with *russus*, an epithet for red (see 'Spectrum'). A more usual and more useful English name for this lovely and easily grown plant is Russian bugloss. The epithet *ruthenicus* is derived from Ruthenia, an old name for an area covering parts of Russia, Ukraine and Belarus. *Centaurea ruthenica* is a pale sulphur-yellow species of knapweed, sometimes known as Russian knapweed, while *Fritillaria ruthenica*, which has faintly chequered, very dark burgundy flowers, is native to Ukraine and southern Russia, where it was first discovered in 1821. The Ruthenian globe thistle, which has vivid blue flowers on top of whitish stems, used to be *Echinops ruthenicus*, but is now reclassified as *Echinops ritro* subsp. *ruthenicus*.

Some regions of Russia have their own specific epithets, among the best known of which are *sibiricus* and (less usual) *sibericus*, designating Siberia. The most familiar Siberian plant is *Iris sibirica*, which in fact has a far larger geographical range than its name suggests and is native to much of Europe as well as Central Asia. Irises are one of the oldest plants in cultivation, and this one was known in England before the end of the sixteenth century. Among other plants identified as Siberian are the tall catmint, *Nepeta sibirica*, the black-stemmed yellow-flowered *Ligularia sibirica*, and several trees, including the Siberian pine, *Pinus sibirica*, and the Siberian fir, *Abies sibirica*. The familiar little violet-blue Siberian squill takes the alternative epithet and is *Scilla siberica*.

The epithet *tataricus* seems to have more or less disappeared along with the old Asian region of Tartary it supposedly indicates, and the range of the Michaelmas daisy *Aster tataricus* includes Korea, Japan and China as well as Siberia. More common is *tauricus*, derived from Taurica, the classical name for the Crimea. *Phlomis taurica* is another good species of this very varied genus, with neat mounds of dark green leaves and bright light-purple flowers, while *Asphodeline taurica* has white rather than the usual yellow flowers. Several subspecies define Crimean plants, such as *Allium flavum* subsp. *tauricum*, with umbels of flowers of an unusual orange colour, although a number of cultivars have flowers that are pink, yellow or cinnamon. Not to be confused with any of the epithets denoting white flowers, such as *albus* and its derivatives, *albanus* refers to the ancient city of Albana in what is now the Russian Republic of Dagestan, between the Caucasus mountains and the Caspian Sea. This region is also known as Caucasian Albania, to differentiate it from the Republic of Albania in Europe, for which the botanical epithet is *albanicus*. Here is an example of where Latin once again trumps English, for *Geranium albanum*, a beautiful species with strongly veined flowers that open pale pink but get darker, and conspicuous lavender-purple anthers, is often called the Albanian cranes-bill, not unnaturally leading many gardeners to imagine it comes from Europe rather than Russia. Also from Caucasian Albania are the dwarf alpine *Androsace albana*, with its dense umbels of very pale pink flowers standing above its curiously folded bright green leaves, and the primrose-yellow pasque flower *Pulsatilla albana*. Plants from the Caucasus more generally are designated *caucasicus*, familiar from the delicately lavender-blue *Scabiosa caucasica*, from which a large number of cultivars have been bred to extend the colour range through different shades of blue, as well as white and cream. *Symphytum caucasicum* is sometimes called blue comfrey and is indeed several shades deeper than *S. azureum*. Another quite well-known epithet, *amurensis*, defining species of berberis, dianthus, linden and other plants, takes its name from the Amur River on the Chinese–Russian border. Perhaps the most striking plants from this region are the white lilac *Syringa amurensis* and a tough little pink called *Dianthus amurensis*, particularly attractive in the violet-flowered cultivar 'Siberian Blue'. Before we leave Russia, we should perhaps mention Sakhalin Island off the east coast of the country, once a penal colony. Outside Russia, its fame rests on the fact that Chekhov published a book about his visit there in 1890, but there are a number of indigenous plants from the island that have the epithet *sachalinensis*, including a fir, a mint and an aster. The most striking plant, however, is the spindle tree *Euonymus sachalinensis*, which was introduced to European cultivation in the 1890s. Unlike the common spindle, *E. europaeus*, which has bright pink fruit, the Sakhalin species has red fruit, but the same contrasting bright orange seeds. It is also valued for the autumn colour of its large leaves. Sometimes listed as if synonymous with the closely related *E. planipes*, it is now more properly named *E. latifolius* var. *sachalinensis*.

The ancient city of Bukhara in what is now Uzbekistan lends its name to the epithet *bucharicus*. The white and yellow spring-flowering *Iris bucharica* has the common name of the Bokhara iris, this old spelling the same as that used for the rugs for which the city is famous. Other plants indigenous to the area include *Fritillaria bucharica*, which has star-shaped greenish white flowers, the bright red *Anemone bucharica*, and *Nigella bucharica*, which has lilac upturned flowers very different from the ordinary love-in-a-mist. Another familiar epithet is *karataviensis*, referring to the Kara Tau (or Black Mountain) range in Kazakhstan. Perhaps the most (to me inexplicably) popular plant from the region is *Allium karataviensis*, a dwarf form with large pink-white globes that only just clear the broad, paired glaucous leaves, because they stand on inelegantly stubby stalks. *Erigeron karataviensis* (syn. *E. alexeenkoi*) is a white chamomile-like daisy often and perhaps unsurprisingly confused with the more usual and much more interesting Mexican fleabane, *Erigeron karvinskianus*, which was named after Baron von Karvinsky, a German plant collector working in South America in the 1840s. The epithet *turkestanica* derives from the old central Asian region of Turkestan, stretching east of the Caspian Sea, bordered by Siberia, Mongolia and Tibet. A well-known plant from this region is *Tulipa turkestanica*, which has white star-shaped flowers opening to reveal a bold yellow splash at their centres, and was discovered and named only in 1873. Equally familiar is the strongly scented biennial clary sage *Salvia turkestanica* (now *S. sclarea* var. *turkestanica*), which was being grown in British gardens as early as 1400 and has distinct lilac bracts.

Not to be confused with Turkestan, Turkey traditionally

straddles Europe and Asia, and has long been a rich source of garden plants, and more particularly of bulbs. The Ottoman Empire, founded in 1299, expanded hugely throughout the reigns of several sultans, and by the time Suleiman the Magnificent was ruler (1520–66) it controlled much of Eastern Europe, including Greece, the Balkans and Hungary, a sizeable portion of western and central Asia extending as far east as the Caspian Sea, parts of what is now the Middle East, and a large swathe of coastal northern Africa. Western plant hunters had travelled through the Empire from the mid-sixteenth century, when the French scholar Pierre Belon made an extended trip, taking in not only Greece and Turkey, but Egypt, Palestine and Syria, writing of his experiences and discoveries in *Les observations de plusieurs singularitez et choses memorables trouvées en Grece, Asie, Judée, Egypte, Arabie, & autres pays etranges* (1553).

'The Dalmatian Cap': *Tulipa*

The plant most linguistically associated with Turkey was the tulip. The *Tulipa* genus owes its name to *dulband*, the Persian word for a turban, the shape of which the tulip's flowers were thought to resemble. Gerard nevertheless describes 'tulip' as 'a Turkish and strange name' and gives the plant the English name 'Dalmatian Cap', taken from a country (now part of Croatia) that the Ottomans had conquered some sixty years before Gerard wrote his book. 'Tulipa groweth wilde in Thracia, Cappadocia and Italy;

in Bizantia about Constantinople, at Tripolis and Alepo in Syria,' Gerard informed his readers. 'They are now common in all gardens of such as affect floures, all over England.' By the time Johnson revised the *Herball*, thirty species were described, but this catalogue would be vastly increased by Parkinson, whose chapter on tulips in *Paradisi* runs to over twenty pages. By then the tulip was called The Turkes Cap, a name we now use of a style of lily, but the *Tulpan* (as Parkinson renders it) was still a style of headgear particularly associated with 'the Turkes of *Dalmatia*'. The yellow-flowered toadflax *Linaria dalmatica* takes its species name from this region, as does the fragrant and light-violet Dalmatian iris, *Iris pallida* var. *dalmatica*.

The epithets for plants that come from what is now modern Turkey also hark back to old regional names: they include *bithynicus*, from the ancient kingdom of Bithynia in north-west Anatolia; *chalcedonicus*, from Chalcedon, a Bithynian town on the coast of the Sea of Marmara; and *byzantinus*, from Byzantium, which is now Istanbul. The spring-flowering Turkish squill is *Scilla bithynica*, and *Fritillaria bithynica* is a distinctive species that has silvered green flowers with lighter lime-green or yellow edges and interiors. Two plants associated with Chalcedon are distinguished by their startlingly red flowers: *Lychnis chalcedonica*, known as Maltese cross because of the shape of its individual flowers, and the shining scarlet turk's-cap lily *Lilium chalcedonicum*. Familiar plants from Byzantium include the soft grey lamb's-ear, *Stachys byzantina*, introduced to Britain in the eighteenth century, and the Byzantine meadow saffron, *Colchicum byzantinum*, a large pink autumn crocus that has been grown in British gardens since the mid-fifteenth century. (The genus name *Colchicum*, incidentally, refers to Colchis, an ancient kingdom

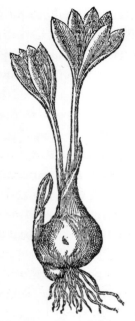

Meadow saffron, *Colchicum*

on the east coast of the Black Sea in what is now Georgia, where it was reported by classical authors that autumn crocuses grew in great numbers.) The bright magenta *Gladiolus byzantinus*, though still listed under this name, is now more properly called *Gladiolus communis* subsp. *byzantinus*.

Plants named after the region of Anatolia take the epithet *anatolicus*, as in the bright yellow *Phlomis anatolica*, and those named after the Turkish capital, Ankara, take the epithet *ancyrensis* (from Ancrya, its name in the ancient world), as in the deep yellow *Crocus ancyrensis* and the light-violet single-flowered pink *Dianthus ancyrensis*. The historical region of Cappadocia was in central Anatolia and is where both the Cappadocian maple, *Acer cappadocicum*, and the blue navelwort, *Omphalodes cappadocica*, come from. The former is usually grown in British gardens as golden ('Aureum') or red ('Rubrum') cultivars, while the latter is usually offered by British nurseries as the cultivar 'Cherry Ingram'. South-west of Cappadocia was Cilicia (now Çukurova), which gives the epithet *cilicius* or *cilicicus*, used of several bulbous species (cyclamen, scilla, colchicum) and an evergreen thyme that has spikes of unusually large light-purple flowers with very prominent anthers.

It is possible that *Cyclamen coum* takes its species name from Koa in eastern Cilicia, although the more usual derivation is from the Aegean island of Kos, from which we also get the Cos lettuce. That said, although the plant has a wide distribution in the wild, from Bulgaria to the Caucasus and from Turkey to Israel, it seems not to grow on Kos; however Alice Coats states that it once did, so perhaps it was reported to be a native there when Philip Miller, first director of the Apothecaries' Garden (later the Chelsea Physic Garden) and author of *The Gardeners Dictionary*, named it in the eighteenth century.

Greece generally is *graecus*, as in the lovely *Fritillaria graeca*, which has bells striped in maroon and green and also grows in the Balkan peninsula. Attica, the ancient region centred on Athens, gives us a specific epithet for the white starry-flowered *Ornithogalum atticum*, and Mount Olympus gives us the huge mullein *Verbascum olympicum*, with its candelabra of yellow flowers, as well as a little pink allium and a low-mounded hypericum. Currently divided between Greece and Turkey, the island of Cyprus is home to the rare *Crocus cyprius*, which has pale lilac petals with dark purple stripes on the back and very striking orange anthers, and was much beloved of E. A. Bowles.

Among epithets for plants originating in Central and Eastern Europe are *carpaticus* or *carpathicus* and *rhodopeus* or *rhodopensis*, from the Carpathian and Rhodope Mountains, and *illyricus*, from the ancient region of Illyria on the Balkan peninsula, best known as the fanciful setting of Shakespeare's *Twelfth Night*. The Carpathian Mountains provide a habitat for a huge number of plants, over a third of all European species. Of these, only a handful refer to their geographical origins in their species names, including a senecio and a ranunculus. One that grows in countless gardens is the vigorously creeping purple bellflower *Campanula carpatica*; while the Carpathian spring snowflake, *Leucojum vernum* var. *carpathicum*, has mustard yellow tips to its petals rather than the usual green.

Anyone who has been to the Rhodope Mountains of Bulgaria will have been astonished by the variety of the region's wild flowers, much helped by old farming practices in which hay is often still cut and gathered by hand and agricultural chemicals are not used. This was once Northern Thrace, where Orpheus descended into the underworld in his ill-fated attempt to rescue his wife Eurydice. One of the plants of the region that has migrated to British gardens is indeed the Orpheus flower, *Haberlea rhodopensis*, which has serrated-edge leaves and light violet-blue flowers on short wiry stems. Among the region's other plants grown in British gardens are the yolk-yellow avens *Geum rhodopeum*, a primrose-yellow scabious *Scabiosa rhodopensis*, and the bright scarlet tulip *Tulipa rhodopea*, which appeared on a Bulgarian postage stamp in 1960. Orsino's Illyria is as rich a source of plants as Orpheus's Thrace, and among its best are *Gladiolus illyricus*, which is not unlike G. *byzantinus* but lacks the white stripes on its magenta petals, and the Illyrian iris, *Iris illyrica*, which has flowers of an intense, deep purple.

The epithet *hispanicus* for Spain has already been mentioned, but the country's regions also provide species names. The north-west region of Asturias gives us both *asturicus* and *asturiensis*, the former for the autumn-flowering, lilac-flowered *Crocus asturicus* (now renamed *C. serotinus* subsp. *salzmannii*), the latter for the dwarf, yellow trumpet-flowered *Narcissus asturiensis*. Other plants take the epithet *valentinus* from Valentia, an ancient Roman colony founded in what is now the port city of Valencia. *Coronilla valentina* is a small shrub with fragrant yellow pea-like flowers, most commonly grown in British gardens in the lemon yellow 'Citrina' cultivar of the *glauca* subspecies. The region is also the source of an autumn-flowering *Leucojum valentinum* (syn. *Acis valentina*) and a little white alpine snapdragon, *Antirrhinum valentinum*. The specific epithet derived from the Balearic Islands off the eastern coast of Spain is familiar from a popular variety of winter-flowering clematis, the fern-leaved *Clematis cirrhosa* var. *balearica*, which has cream-coloured flowers, the interiors of which are lightly speckled with maroon. The mat-forming, white-flowered rockery plant *Arenaria balearica*, which has been grown in British gardens since 1787, is sometimes called Balearic pearlwort, but is also known as Corsican sandwort, being also native to that island, and to the western Mediterranean more generally. Both an ivy and a salvia native to the Canary Islands take the specific epithet *canariensis*, but perhaps the most distinguished garden plant from the islands is the fern-leaved and tender *Lavandula canariensis*, named by Philip Miller in 1768, which has flowers in a candelabra arrangement rather than the more usual single spike.

The Iberian peninsula comprises both Spain and Portugal, and while the epithet *ibericus* can relate to this region (as in the species of scorpion *Buthus ibericus*), in botany it more often refers to the Georgian kingdom of Kartli, which was known as Iberia in the ancient world. So the dwarf *Iris iberica* is correctly known as the Georgian iris and that garden favourite *Geranium ibericum* is the Caucasian cranesbill. Some species of the *Iberis* genus are native to the Iberian peninsula, but the English name of candytuft refers to Candia, an ancient name for Crete, which is where Gerard acquired seeds of the plant.

The specific epithet for Portugal, *lusitanicus*, is another of those that looks back to the ancient world, in this case to when most of the present-day country and part of Spain became the Roman province of Lusitania. The Portugal laurel, *Prunus lusitanica*, was introduced to England in the mid-seventeenth century, when a specimen was grown at the Oxford Botanic

Garden. It became popular because it was thought hardier than the cherry laurel *P. laurocerasus*, which had been introduced some seventy years earlier and was pessimistically considered tender, grown in tubs or pots that could be moved in under glass for the winter. Unlike some French lavenders, the Portuguese subspecies, *Lavandula pedunculata* subsp. *lusitanica*, is reputedly hardy, and has the 'ears' familiar from *L. stoechas*. There is also a laurel from the Portuguese archipelago of the Azores, *Laurus azorica*, which the RHS lists as Canary Island laurel, although very similar plants found growing on those islands now have their own species name, *L. novocanariensis*. Probably the best-known plant from the Portuguese island of Madeira is the huge but tender and short-lived *Geranium maderense*, the largest of the genus. Accustomed to rather better weather than British gardens can provide, these plants hate both the wet and the cold. If they thrive they take up a great deal of space, able to reach a spread of two metres and a height of around one and a half metres. They have handsome fern-like foliage, which is just as well since they tend to take a couple of years to produce their bright magenta-pink flowers and then die from the effort. Also large is the bright yellow Madeira broom, *Genista maderensis*, but the Madeira marguerite, *Argyranthemum maderense*, is a compact shrub with primrose-yellow daisy flowers, and despite its name is in fact native to nearby Lanzarote and Fuertaventura.

The specific epithet for Italy is, naturally enough, *italica*, so quite why Linnaeus applied it to the Balearic sage, *Phlomis italica*, a species with soft pink flowers and greyish foliage that is native to Mallorca and Menorca, is something of a mystery. The fenugreek-scented curry plant, *Helichrysum italicum*, may have the Italian species name but in the wild is distributed around the whole of the Mediterranean, while the

Italian arum, *Arum italicum*, usually grown in gardens in one of its marble-leaved cultivars, has an even wider distribution. As with Spain, individual regions of Italy have their specific epithets. A stylised iris has been the symbol of Florence for centuries, and although the heraldic flower is crimson, 'dyed red by division', according to Dante in reference to the city's bloody history, *Iris florentina* is a pure unspotted white. Doubt has been cast upon the naming of this plant, which is now sometimes regarded as a variety of *I. germanica*, which might be thought unfortunate given the much resented year-long occupation of the Italian city by German forces during the Second World War, at the end of which many historic bridges and buildings were damaged or destroyed. Tuscany more generally takes an epithet derived from its ancient Etruscan culture, the spring-flowering *Crocus etruscus* having buff petals streaked with purple.

Of the epithets for other European countries, *gallicus* will be familiar both as the source of the adjective used of our French neighbours and because of *Rosa gallica*, which grows wild in several other parts of Europe and is the only red rose to do so. It is possible that *R. gallica* is the oldest rose in cultivation, and it may well have been the Apothecary's Rose (*R. gallica* var. *officinalis*) that Edmund of Lancaster brought to Britain in around 1279. We now refer to Gallica roses as a group, amongst which is another very old rose, the raspberry-rippled Rosa Mundi (*R. gallica* var. *versicolor*), associated with Fair Rosamund, the mistress of Henry II, though not entering the written record until over 400 years after her death. The name Rosamund or Rosamond can mean rose of the world (*rosa mundi*) or pure rose (*rosa munda*), but originally derives rather less romantically from an Old German name meaning 'protector of horses'.

Germany is indicated by the epithet *germanicus*, and while

a great many plants are native to that country, the bearded iris
Iris germanica has a very wide distribution right across Europe
to Iran and Nepal. It has been in cultivation for a very long
time, and although it was recorded growing in the garden of
a German abbey in the ninth century, it seems that Linnaeus
gave it the species name *germanica* because the specimen he
had been sent was from another German garden. Now considered a hybrid, it has been renamed *Iris × germanica*. There
are records of the medlar, *Mespilus germanica*, being cultivated
in 7 BC. Although the tree was introduced to Britain by the
Romans, and frequently found in monastery gardens during
the medieval period, the eighteenth-century Oxford professor
of botany John Sibthorp believed it to be native to Germany.
It is not, but grows wild across a large area, stretching from
the south-eastern Balkan peninsula to south-western Asia.
However, Sibthorp was one of Linnaeus's correspondents,
which is probably why the the plant got its Germanic specific
epithet in *Species Plantarum*.

The British Isles have their own specific epithets for their
constituent parts. There ought to be an English rose, a Welsh
daffodil, a Scottish thistle and an Irish shamrock exemplifying
these species names, but while *Narcissus cambricus* is one of the
many synonyms for the widely distributed *N. pseudonarcissus*,
you will look in vain for a '*Rosa anglica*', a '*Cirsium scoticum*' or
a '*Trifolium hibernicum*'. English stonecrop, however, is *Sedum
anglicum* and there is also an Anglicum group of the monkshood *Aconitum napellus* subsp. *napellus*, which is sometimes
offered by nurseries as *A. anglicum*. The ubiquitous but lovely
lemon-yellow Welsh poppy is *Meconopsis cambrica*, and while
the fern *Polypodium cambricum* is available in a number of cultivars, the one cotoneaster native to Britain, *C. cambricus*, grows
only on Great Orme Head near Llandudno in North Wales,

is critically endangered, and is not considered gardenworthy. The tiny Scottish primrose, *Primula scotica*, also grows in the wild in quite a small area that includes Orkney, Caithness and Sutherland, but its umbels of fragrant, bright purple flowers with a yellow eye have made it a popular alpine for rockeries and troughs. Scots lovage, *Ligusticum scoticum*, is an attractive and edible umbellifer with off-white flower-heads and green leaves that sometimes have purple margins. Crossing the sea we come to Ireland and find the vigorous Irish or Atlantic ivy, *Hedera hibernica*, often used in gardens to provide rapid but elegant cover for unsightly buildings or structures, and the Irish juniper, *Juniperus communis* 'Hibernica', which has the same fastigiate habit as the Irish yew.

Before we leave Europe, we should mention the somewhat vague epithet for the entire continent, *europaeus*. This may be familiar from the olive tree, *Olea europaea*, presumably given this name by Linnaeus because it had long been cultivated on the European mainland, though it also grows in much of Asia. The common lime tree, *Tilia* × *europaea*, which I am lucky enough to have lining the street where I live in London, is, however, truly European, a naturally occurring hybrid between two other limes, *T. cordata* and *T. platyphyllos*. A popular garden plant, of which there are a number of cultivars, is the yellow globeflower, *Trollius europaeus*, native to Britain as well as mainland Europe and referred to by Gerard as the Globe Crow-foot. Its genus name is derived from the German *Trolleblume*, itself a contraction of *die rolle Blume* or roll flower, a reference to the way in which the petals are rolled or closed into the sphere that gives the plant its English name.

Gardeners such as the Tradescants took advantage of the opening up of the 'New World' of the Americas in the early seventeenth century, when British settlers established colonies first in Virginia and then in New England. The elder Tradescant had imported and started growing American plants by 1634, listing some forty of them in the back of his copy of Parkinson's *Paradisi*, and his son made the first of his three plant-hunting expeditions to North America three years later. By this time the Virginia creeper, *Parthenocissus quinquefolia* (its genus name from the Greek *parthenos* = virgin + *kissos* = ivy), and what Parkinson called the Virginia sumach, *Rhus typhina*, were already being grown in Britain – the latter 'onely kept as a rarity and ornament to a Garden or Orchard', no one having yet 'made any tryall of the Physicall properties'. The younger Tradescant brought back a number of plants from his trips, perhaps including the sensitive plant, *Mimosa pudica*, collected en route in Barbados, and the Michaelmas daisy that bears his name, *Aster tradescantii* – or at any rate did so until the RHS declared it 'misapplied'. He also brought back the spiderwort that would be named to honour both him and his father, *Tradescantia virginiana*. (In one of those regular botanical reshuffles, this species was renamed *Tradescantia* × *andersoniana* to honour the American botanist Edgar Anderson, who had written a book on the genus in 1935.) Other Virginian plants that followed tradescantias into British cultivation in the seventeenth century include the blue Virginia cowslip, *Mertensia virginica* (syn. *M. pulmonarioides*, because it indeed looks like a lungwort, though its leaves are smooth), and the swamp magnolia, *Magnolia virginiana*. Both these plants were introduced to Britain by John Banister (1654–92), sent to America by Henry Compton, Bishop of London, who took full horticultural advantage of the fact that he was also responsible for religious affairs in the new American colonies. Compton filled his episcopal garden at Fulham Palace with 'a greater variety of curious exotic plants and trees than had at that time been collected in any garden in England'. A number of these had been collected by Banister, who would also provide John Ray with a list of plants he had seen in Virginia, which when published in the second, 1688 volume of Ray's *Historia Plantarum* became the earliest catalogue of American flora. Other American plants were given the more general species name *americanus*, including *Ceanothus americanus*, also introduced by Banister. *Callicarpa americana* was introduced to Britain in the early eighteenth century by the well-travelled Mark Catesby, author and for the most part illustrator of the lavish two-volume *Natural History of Carolina, Florida and the Bahama Islands* (1729–47). The shrub takes its genus name from Greek *kalos* = beautiful + *karpos* = fruit, and is indeed grown for its startlingly bright-purple berries. The specific epithet *americanus* sometimes extends its geographical range to other parts of the Americas: the century plant, *Agave americana*, for example, is a native of Mexico, introduced to Britain in 1814.

Plants from other regions of North America had epithets created for them, notably Florida and Canada. The southern sugar maple is *Acer floridanum* and the Florida yew *Taxus floridana*. The evergreen shrub *Illicium floridanum* is sometimes known as purple anise because of the scent its leaves emit when crushed, but the plant is highly toxic. Despite this, and despite intriguing maroon flowers that smell of fish if you get too close to them, the plant is available in several cultivars with different-coloured flowers and variegated foliage. *Floridanus* should not be confused with *floridus*, which means showy or free-flowering, nor *canadensis*, meaning from Canada or the American north, with *candens*, which means a shining white. This might prove particularly confusing in the case of the Canadian burnet, *Sanguisorba canadensis*, which has white rather than the usual dark red flowers. The serviceberry *Amelanchier canadensis* also has a mass of white flowers, which stand out against young foliage of a deep bronze, and is native to both Canada and eastern states of the USA, as is the eastern redbud, *Cercis canadensis*, of which 'Forest Pansy' is a particularly beautiful cultivar with bright pink flowers followed by dark purple heart-shaped leaves. A couple of American states have become a little less recognisable when Latinised – Louisiana (*ludovicianus*) and New England (*novae-angliae*) – but the rest are easy to decode, and include *missouriensis* (Missouri), *californicus* (California), *texensis* (Texas) and *oreganus, oregonus* and *oregonensis* (Oregon and the north-west more generally). California and Texas in particular have been the source of a number of plants grown in British gardens. *Fremontodendron californicum* is a large shrub with deep-yellow flowers that requires a sheltered position here, and *Zauschneria californica*, sometimes known as the Californian fuchsia, provides a vivid splash of scarlet to the autumn border. The California poppy, *Eschscholzia californica*, with its feathery green foliage and cupped flowers, is now available in a wide range of colours, from dark red to creamy white, but none is as beautiful as the bright yellow-orange original that grew in such vast sheets in California that it is sometimes said to be the origin of the term 'the Golden West', apparently coined well before the gold-rush of the mid-nineteenth century. Great swathes of the Texas bluebonnet, *Lupinus texensis*, are a familiar sight in the state and among the native plants Lady Bird Johnson, widow of President Johnson, campaigned to preserve and reintroduce alongside highways and elsewhere. In British gardens *Clematis texensis* is a popular, because modestly sized, species grown in a large number of cultivars which have either thick-petalled and urn-shaped flowers or more open trumpet-shaped ones in a wide range of colours.

It was in the Caribbean that Hans Sloane (1660–1753) first made his reputation as one of the world's great collectors, not only of plants but of ethnographical artefacts and other objects. He sailed to Jamaica in 1687 as the personal physician to the Duke of Albermarle and spent fifteen months on the island, returning to England with a long and detailed journal and hundreds of specimens of plants, animals and 'curiosities', all carefully labelled. He would subsequently publish both a Latin dictionary of Jamaican plants (1696) and a two-volume *Natural History of Jamaica* (1707–25), which Linnaeus greatly admired and made use of while writing his *Species Plantarum*. Given the difference in climate between the West Indies and Britain, plants indigenous to the islands are grown here, if at all, as hothouse exotics rather than in gardens. The generic epithet for the region, *caribaeus* (with only one b), is used for such plants as the lobster claw *Heliconia caribaea*, and the spider lily *Hymenocallis caribaea*. Individual countries have their own epithets, as in the Jamaican orchid *Bulbophyllum jamaicense*. The yellow walking iris is *Trimezia martinicensis*, referring to its native habitat of Martinique, and *Aloe vera* has the synonym *A. barbadensis* because it is a native of Barbados.

The wealth of plants from South America were first reported by Nicolás Monardes (after whom Linnaeus named the bergamot genus *Monarda*), in his *Historia medicinal de las cosas que se traen de nuestras Indias Occidentales*, published in three volumes between 1565 and 1574. Monardes was the personal physician of Philip II of Spain and had not in fact been to South America, but his sons Jerónimo and Dionisio had, and his book is an account of the plants they had found there and of their possible medical benefits. It was translated into English by John Frampton in 1577 as *Joyful News out of the New Found World*, and much of the rejoicing was to do with the sheer range and novelty of the plants it described, not simply their potential medical application. It was the first time, for example, that the sunflower had been described, and the plant's principal attraction was that it 'showeth marveilous faire in Gardens', which soon led to it becoming a favourite 'exotic' among British gardeners, partly because of its impressive size. Gerard boasted that one he had raised from seed had 'risen vp to the height of fourteene foot, where one floure was in weight three pounds and two ounces, and crosse ouerthwart

Gerard's 'greater Sun-floure',
Helianthus annuus

the floure by measure sixteene inches broad'. He called it 'the golden floure of Peru', and other garden plants from that country have *peruvianus* as a species name. The bright red, low-growing *Verbena peruviana* is one of the parents of several garden hybrids in different colours, and *Physalis peruviana* is the cape gooseberry, the fruit of which is eaten both fresh and dried. *Scilla peruviana*, however, despite the name given to it by Linnaeus, is of Mediterranean origin and is more accurately known as the Portuguese squill. Equally confusing is the botanical name for the marvel of Peru, *Mirabilis jalapa*, Jalapa being a city in Mexico, while the plant is a native of tropical South America more generally.

The flaming red *Begonia boliviensis* was introduced to Europe from the Bolivian Andes in the 1860s, and *andinus* is the specific epithet for plants native to those mountains, such as the Chilean fern *Polystichum andinum*. Chile has its own epithet, *chilensis*, and *Libertia chilensis* is one of several of this genus (including *L. formosa*) that originate there. A much larger plant that is now grown in British gardens is the Chilean wine palm, *Jubaea chilensis*, hardy enough to survive our winters. *Victoria amazonica* is the largest of all water-lilies, one of the natural marvels at the Great Exhibition of 1851, at which children stood on its pads, and it owes its species name to the Amazon River. Less easy to guess at is another riverine epithet, *fluminensis*, which refers to Rio de Janeiro. The reason for this is that when the Portuguese explorer Gaspar de Lemos arrived in Brazil, he assumed that Guanabara Bay was the mouth of a river, and since he made landfall on 1 January 1502 he gave the place a name meaning 'January River'. The Latin for this is *Flumen Ianuarius*, hence the epithet *fluminensis*, as in the ground-hugging, white-flowered spiderwort, *Tradescantia fluminensis*, particularly popular in such variegated cultivars as 'Quicksilver' and 'Maiden's Blush'. The species name of the tall, airy *Verbena bonariensis*, which has been grown in English gardens since 1726, may also be not immediately obvious, but is derived from Buenos Aires. The plant has the less common synonym of *V. patagonica*, suggesting a habitat further south than the Argentine capital. Other plants native to the southern tip of South America sometimes take the epithet *magellanicus*, after the Magellan Straits or the Portuguese explorer who named them, as in *Fuchsia magellanica*, which is native to Argentina and Chile.

Of the countries of Central America, the one that has been the greatest source of plants for British gardens is Mexico, home of the dahlia, the zinnia and the ageratum. Other native plants include the purple-flowered *Agastache mexicana*, *Philadelphus mexicanus,* which has cream-coloured cup-shaped flowers, and the bright lime-yellow Mexican stonecrop, *Sedum mexicanum.*

The first plants from the Antipodes reached Britain in 1701, brought back by the adventurer William Dampier and catalogued in Ray's *Historia Plantarum.* A more scientific enterprise than Dampier's was undertaken by Joseph Banks, who funded and led a group of botanists accompanying Captain Cook on the *Endeavour,* which left England in 1768 with the principal aim of observing the Transit of Venus in Tahiti. When the *Endeavour* reached Australia in April 1770, landing in what would be named New South Wales, Banks and his team went off in search of plants, collecting and drying specimens to take back to England. Such was the number of plants they found there that Cook named the place where they had landed Botany Bay. Australia was known at this time as New Holland or Terra Australis, the Latin for Southern Land. Indeed, the specific epithet *australis* means 'southern' rather than from Australia in particular, for which the epithet is the rarer *australiensis,* most often used for plants that are not grown outside the continent, such as the Australian cashew, *Semecarpus australiensis* or the northern Australian wild rice *Oryza australiensis. Australis* is, however, sometimes used for Australian species of plants that also grow in other parts of the world, such as Australian indigo, *Indigofera australis.* As already noted, the epithet *austrinus* similarly means southern – as in the Florida flame azalea, *Rhododendron austrinum,* which in fact grows wild in most of the southern states of the USA – but it is also applied to some Antipodean plants, such as the yellow-flowered *Hibiscus austrinus,* which is native to Western Australia.

And this is where we come to rest after circumnavigating the globe. Inevitably we have been unable to call in at a large number of countries, but this whistle-stop tour and the word lists below give some idea of how plants from around the world have reached these shores, been coaxed into growing in an often very different climate from their native one, enriching our gardens and becoming so familiar that without their species names we might forget just how far many of them have travelled.

Word Lists

GENERAL

australis = southern (or Australian)

austrinus = southern

barbarus = foreign

borealis = northern

occidentalis = western

orientalis = eastern

HABITAT

acraeus = growing at high altitudes

agrarius, agrestis = growing in fields

algidus = from high mountains

alpestris = alpine, below the timber line

alpinus = alpine, above the timber line

ammophilus = sand-loving

amphibius = growing on land or water

apricus = sun-loving or open to the sun

arenarius = growing in sandy places

aridus = favouring dry places

arvensis = growing in cultivated fields

bufonius = growing in damp places

campestris = of the fields

clivorus = of the hills

collinus = of the hills

maritimus = growing by the sea, coastal

montanus = of mountains

nemorosus = growing in woodland

palustris = growing in marshes

pratensis = of meadows

rivalis, rivularis = growing by streams or rivers

rupestris = growing on rocks

rupifragus = 'rock breaking', i.e. growing among rocks

saxatilis = growing among rocks

sylvaticus, sylvestris = growing in woods or forests

uliginosus = growing in bogs or waterlogged ground

PLACES, BY GEOGRAPHICAL REGION

EUROPE AND NEAR EAST

aegeus = the Aegean

aetnensis = Mount Etna

albanensis = St Albans

albanicus = Albania

algarvensis = the Algarve

anatolicus = Anatolia

ancyrensis = Ankara

anglicus = England

asturicus, asturiensis = Asturia, Spain

atticus = Attica

austriacus = Austria

azoricus = the Azores

balearicus = the Balearic islands

bavaricus = Bavaria

baeticus = Andalusia

belgicus = Belgium

bithynicus = Bithynia on the southern shore of the Black Sea

byzantinus = Byzantium, now Istanbul

cambricus = Wales

canariensis = Canary Isles

cantabricus = Cantabria, Spain

cantabrigiensis = Cambridge, England

cantianus = Kent

cantuariensis = Canterbury

cappadocicus = Cappadocia, Anatolia

carpaticus, carpathicus = the Carpathians

chalcedonicus = Chalcedon, Anatolia

cilicicus = Cilicia, now Çukurova in Turkey

cismontanus = 'on this side of the mountain', the southern, Roman side of the Alps

corsicus = Corsica

cous = Kos, or Koa in Turkey

creticus = Crete

croaticus = Croatia

cyprius = Cyprus

danicus = Denmark

etruscus = Tuscany

europaeus = Europe

florentinus = Florence

gallicus = France
germanicus = Germany
graecus = Greece
helveticus – Switzerland
hibernicus = Ireland
hispanicus = Spain
hyrcanus, hircanicum =
    Hyrcania, ancient region
    between the Caspian Sea
    and the Alborz Mountains
    in Iran
ibericus = the Iberian
    peninsula or the Georgian
    Caucasus
illyricus = Illyria, western
    Balkan peninsula,
    Adriatic coast
italicus = Italy
lusitanicus = Portugal
monspeliensis = Montpellier
olympicus = Mount Olympus
oxonianus = Oxford
ponticus = Pontus, south of
    Black Sea (Turkey)
pyrenaeus, pyrenaicus = the
    Pyrenees
rhodopensis, rhodopeus =
    Rhodope Mountains,
    Bulgaria
sabatius = Savona on the
    Ligurian coast
scoticus = Scotland

siculus = Sicily
stoechas = the Stoechades,
    now Îles d'Hyères, France
suecicus = Sweden
tarentinus = Taranto, Italy
uplandicus = Uppland, Sweden
valentinus = Valencia, Spain

AFRICA

abyssinicus = Ethiopia
aethiopicus = (South) Africa
afer, afra = African (usually
    North African coast)
africanus = Africa
algeriensis = Algeria
atlanticus = the Atlas
    Mountains
capensis = the South African
    Cape
madagascariensis =
    Madagascar
tingitanus = Tangier

RUSSIA AND CENTRAL ASIA

albanus = Caucasian Albania,
    now Dagestan
aleuticus = Aleutian Islands
altaiacus = Altai Mountains in
    central and eastern Asia
amurensis = Amur River on
    Chinese–Russian border
armeniaca = Armenia

bucharicus = Bukhara, Uzbekistan

caucasicus = Caucasus mountains

ibericus = the Georgian Caucasus or the Iberian peninsula

karataviensis, karatavicus = Kara Tau mountain range, Kazakhstan

russicus = Russia

ruthenicus = Ruthenia, and more generally Russia

sachalinensis = Sakhalin, Russia

sibericus, sibiricus = Siberia

tataricus = Tartary, Russia

tauricus = Crimea

turkestanicus = Turkestan

INDIAN SUBCONTINENT

bengalensis, benghalensis = Bengal

borbonicus = Île Bourbon (now Réunion) in the Indian Ocean

cachemirianus, cachemiricus, cashmerianus, cashmirianus, kashmirianus = Kashmir

emodi, emodensis = Himalayas

himalaicus, himalayensis = Himalayas

indicus = India

napaulensis, nepalensis = Nepal

sikkimensis = Sikkim

thibetanus, thibeticus, tibetanus, tibeticus = Tibet

FAR EAST

chinensis = China

coreanus, koreanus, koraiensis = Korea

formosanus = Formosa, now Taiwan

hupehensis = Hupeh (now Hubei) province, China

japonicus = Japan

mandschuricus, mandshuricus, manchuriensis = Manchuria, China

niponicus, nipponicus = Japan

pekinensis = Peking, now Beijing

sinensis = China

yedoensis, yesoensis, yezoensis = Tokyo, but more generally Japan

yunnanensis = Yunnan Province, China

MIDDLE EAST

aegypticus = Egypt

aleppicus = Syria

alexandrinus = Alexandria

libani, libanotis, libanoticus =
    Lebanon

persicus = Persia, now Iran

syriacus = Syria

CENTRAL AND
SOUTH AMERICA AND
THE CARIBBEAN

amazonicus = Amazon River

andinus = Andes

bahamensis = the Bahamas

barbadensis = Barbados

bermudianus = Bermuda

boliviensis = Bolivia

bonariensis = Buenos Aires,
    Argentina

brasilianus, brasiliensis = Brazil

caribaeus = Caribbean

chilensis = Chile

cubensis = Cuba

fluminensis = Rio de Janeiro

jamaicensis = Jamaica

magellanicus = southern tip of
    South America

martinicensis = Martinique

mexicanus = Mexico

patagonicus = Patagonia

paraguayensis = Paraguay

peruvianus = Peru

NORTH AMERICA

americanus = America
    generally

californicus = California

canadensis = Canada or north-
    eastern America

floridanus = Florida

missouriensis = Missouri

novae-angliae = New England

novi-belgii = New York

oreganus, oregonensis, oregonus
    = Oregon and the
    north-west

sitchensis = Sitka, Alaska

texensis = Texas

virginicus, virginianus =
    Virginia

AUSTRALIA AND
SOUTH SEAS

australiensis = Australia

australis = southern, but also
    Australia

austrinus = southern, but also
    Australia

novae-zelandiae = New
    Zealand

POLAR

arcticus and *antarcticus*

ANATOMY

Given that plants were first studied and categorised for their supposed medicinal value, it is unsurprising that some of them are named after parts of the body. Although many of them do indeed have healing qualities, and are still used in drugs today, one of the silliest notions in early botany and medicine was the so-called doctrine of signatures. This theory, which stretches way back to Dioscorides and was still prevalent enough for John Ray to feel the need to refute it some 1600 years later, was that an individual part of the body could be healed by plants that had features such as seed, leaf or root that resembled it. An originally pagan belief was adopted by Christians on the grounds that the doctrine was an indication of God's benevolence in pointing humans towards the curing of their ills. The idea was given particular currency by Philippus Aureolus Theophrastus Bombastus von Hohenheim, otherwise known as Paracelsus (1493/4–1541), an unfortunately influential philosopher and alchemist of whom perhaps the best that can be said is that he inspired an early verse drama by Robert Browning. Despite Rembert Dodoens rejecting the doctrine of signatures in 1583, when he observed that it 'has received the authority of no ancient writer who is held in any esteem', Robert Turner, the seventeenth-century botanist, astrologer and author of *Botanologia: The Brittish Physician: or, The Nature and Vertues of English Plants* (1664), would assert that: 'God hath imprinted upon the Plants, Herbs, and Flowers, as it were in Hieroglyphicks, the very signature of their Vertues.' William Coles was another botanist who believed this long-exploded theory, writing for example in his *Adam in Eden: or, natures paradise* (1657) that: 'The down of quinces doth in some sort resemble the hair of the head, the decoction whereof is very effectual for the restoring of hair that is fallen off by the French pox.' In fact the down on the quince looks nothing like hair, not even the first growth on a baby's head, and since it rubs off very easily it hardly inspires confidence as a remedy for successful and permanent regrowth. Ray was having none of this nonsense, writing in his Cambridge Flora: 'Of the plants specifically said to be appropriate to a particular part of the body or to a disease far the greater number have no signature [...]. Different parts of the same plant such as leaves, roots, flowers and seeds, have signatures not merely different but contradictory [...] Parts of some plants represent parts of the body with which they violently disagree. Thus the fruit of Anacarditum represents a heart and is nevertheless poisonous; the juice of the spurges is like milk but no one is such an imbecile as to give it to a nursing mother; the flesh of Mespilus [medlar] is like human excrement and similar in colour, but is not suitable as an aperient [to relieve constipation].'

This does not alter the fact that the tropical plant *Anacardium* (as it is now spelled) takes its botanical name from the heart-shaped seed, which is located not at the centre of the fruit but grows at one end of it. The name is from the Greek *ana*, meaning up + *kardia* = heart: for the prefix, think of Anacapri, that

part of Capri that is situated on a higher elevation than the rest
of the island; for the second part, think of cardiology.

It is with the heart that we begin our botanical exploration
of the body. As its English name implies, common motherwort
was once widely used for treating women's medical complaints; but it was additionally thought by Parkinson 'to bee of
much use for the trembling of the heart', and has the botanical
name *Leonurus cardiaca*. In Parkinson's time it was not merely
the wild *Viola tricolor* that was called heartsease, but also some
of the cultivated violas or pansies, but these names had less to
do with cardiology than with the heart as a symbol of love.
'Pansy' is a corruption of the French word for the plant, *pensée*
or thought, and pansies were an emblem of thinking about
or remembering the person you loved – a notion echoed in
the plant's name in other languages, such as Italian (*viola del
pensiero*) and Spanish (*pensamiento*). Violas and pansies in fact
have heart-shaped leaves, which is to say they resemble the universal symbol for the heart rather than the organ itself, and this
is the most usual meaning of heart-related names in botanical
Latin. The giant Himalayan lily, *Cardiocrinum giganteum*, for

'The white garden Violet', *Viola blanda*

example, takes its genus
name from its huge, heart-
like leaves. By the same
token, the campion *Silene
cardiopetala* was given
this name because of its
flowers' heart-shaped
petals. Other plants with
heart-shaped leaves go to
the Latin word for our
beating centre, *cor* (gen.
cordis), to produce the

epithets *cordatus* and *cordifolius*. While the leaves of *Houttuynia cordata* form perfect green hearts, those of the pickerel weed, *Pontederia cordata*, are somewhat attenuated. The leaves of *Crambe cordifolia* suggest a heart that has been worn rather ragged, though they certainly distinguish the species from seakale, *C. maritima*, which has glaucous foliage very similar to the kale now sold in supermarket packages.

The prefix *cephala-* may be vaguely familiar from 'cephalopods', the name used for members of the squid and octopus family, or even (to those who like obscure words) from 'dolichocephalic', meaning having a long skull or face and not much used outside the works of Elizabeth Bowen. *Cephalus* is the Latinised version of the Greek word for head, *kephale* (and should not be confused with *cephalonicus*, which denotes a plant originating in the Greek island of Kephalonia). Anyone who has seen the large primrose heads of the giant scabious, *Cephalaria gigantea* (syn. *C. tatarica*), standing tall above their serrated green leaves will see the point. Similarly it is the large, bright yellow thistle-heads of *Centaurea macrocephala* that give the plant its botanical name, the prefix *macro-* in botany usually meaning large even though it is derived from the Greek *makros*, which means long. Logically, the specific epithet should be *megacephalus*, since *megas* is the Greek for large, but this species name appears to be restricted to animals. Other plants with smaller heads of flowers than is usual in a genus have the species name *microcephalus*, *micro-* (from the Greek *mikros*) meaning small. For gardeners, this is hardly enticing, as may be judged by comparing the blazing star most people grow, *Liatris spicata*, with the sparsely flowered *L. microcephala*, which unsurprisingly is no longer listed in the RHS Plant Finder. The Latin word for head is *caput*, which gives us the specific epithet *capitatus*, referring to plants with flowers growing in a dense

head, such as the unusually floriferous *Thymus capitatus*, the
stems of which end in packed clusters of purplish pink flowers,
or the Himalayan primrose *Primula capitata*, which has spherical 'drumstick' heads of massed purple flowers.

Another primula, *P. auricula*, takes its name from the shape
of its leaves, *auris* and *auricula* being Latin words for the ear.
It is not clear who first called this plant *Auricula ursi*, which
resulted in the old English name of Beares eares, or indeed
whether this person had actually seen a bear, and if so what
it was about the plant's leaves that suggested this particular
animal. An alternative and rather more accurate name for the
auricula in Gerard's time was the mountain cowslip, since its
natural habitat is the mountains of central Europe. Early botanists also dubbed it *Sanicula alpina* because of what Gerard
called its 'singular facultie in healing of wounds, both inward
and outward'. This genus name was taken from *sanitas*, the
Latin word for health, but the auricula is not related to the
wood sanicle, *Sanicula europea*, which was also once supposed
to cure a wide variety of ills.

Depending on who you believe, *auriculatus* can either
mean ear-shaped or earlobe-shaped, though given the wide
variety of earlobe shapes seen in humans it is unclear what the
latter definition means. As in auriculas, the epithet is usually
applied to leaf shape, but seems to be so vague as to be more
or less useless as a description. The definition given by some
exasperated patient to the term 'phallic symbol' (something
that is longer than it is wide) might equally apply at first sight
to the leaves of a number of plants bearing the species name
auriculatus, such as the African evergreen *Buddleja auriculata*
or *Rhododendron auriculatum*. At least the egg-yolk-orange
Coreopsis auriculata has leaves quite different in form from the
pinnate or deeply lobed ones of other species. However, the

botanical term auriculate more often refers to what the RHS's indispensable *Index of Garden Plants* (1994) calls 'an ear-like lobe or outgrowth, often at the base' of a leaf, which anyone peering closely at the aforementioned rhododendron might notice – though what really distinguishes the species to the gardener's eye are its huge white flowers. B. Daydon Jackson's *Glossary of Botanic Terms* (1900) supplies examples of auriculate leaves: the orange tree and the sage – though neither genus has an *auriculatus* species currently listed. And it is no good looking at the leaves of the Indian *Senna auriculata*, which are pinnate; but its bright yellow flowers have petals that might be described as ear-shaped if your idea of an ear is the kind sported by a toy mouse. Speaking of which, the name of the forget-me-not genus, *Myosotis*, is the Greek word for 'mouse-ear', referring to the shape of the plant's leaves.

Noses are slightly easier to deal with than ears, because plant names taken from the Latin *nasus* (which gives us 'nasal') tend to refer not to the shape of leaf or flower but to the effect these plants have upon the nostrils. What we call nasturtiums derive what was also once their botanical name, *Nasturtium indicum*, from *nasus* + *tortus* = twisted, because the pungent taste of the edible flowers makes the nose twitch. The old species name, echoed in the old common one of Indian Cresse, refers not to India but to the plant's native habitat of Central and South America, which were sometimes known as the Indies in the sixteenth century, just as the Caribbean islands are still known as the West Indies. Linnaeus replaced the nasturtium's wholly appropriate and enchanting original name with *Tropaeolum majus*, the genus name taken from *tropaeum*, the Latin word for trophy. His somewhat fanciful reason for this new name (which is also fiendish to spell) was that the round-shaped leaves and the orange flowers reminded

'Indian Cresse' or nasturtium,
Tropaeolum majus

him of the captured shields and bloodied helmets of defeated armies that the Romans traditionally displayed in victory parades. *Majus* simply means bigger or major, referring to the much larger flowers of the nasturtium compared with other species of *Tropaeolum* such as the canary creeper, *T. peregrinum*, and the flame creeper, *T. speciosum*.

The even more pungent watercress, which is in the same order but not the same family, used to bear the splendidly triple-barrelled name of *Rorippa nasturtium-aquaticum*. I had always imagined that the genus name was derived from the Latin *ripa*, meaning riverbank, as in the adjective riparian, but it is in fact a Latinised version of *rorippen*, an Old Saxon word for members of the cress family. In any case, some spoilsport has renamed watercress *Nasturtium officinale*, which at least rescues the otherwise discarded nose-twisting name. The equally delicious land cress, incidentally, is *Barbarea verna*, its genus name taken not from *barba*, the Latin word for that other facial feature, a beard, but a contraction of *Herba Sanctae Barbarae* (St Barbara's herb), the name given to it by Dodoens – though quite why the plant should belong to a third-century Greek martyr beheaded by her own father is a mystery. The other nose-tickling plant often found in gardens is *Achillea ptarmica*, the species name of which was adapted from Dioscorides' *ptarmikos*, an onomatopoeic word meaning 'causing to sneeze'. The plant has the common name of sneezewort, though was used by early physicians chiefly to cure toothache.

Those botanical names that do refer to the shape of the nose are usually derived from the Greek word for it, *rhis* (gen. *rhinos*), – hence the rhinoceros (*keras* = horn) or rhinoplasty (*plassein* = to mould), which is what nose-jobs are called in Beverly Hills. Yellow rattle is *Rhinanthus minor*, meaning 'little nose-flower', because of the beaky upper lip of its yellow flower. The addition of *anti-*, meaning counterfeiting, produces *Antirrhinum*, since their flowers were thought to resemble animals' snouts. They are now more commonly known as snapdragons, a name for children who enjoy squeezing the lobed flowers to make their 'mouths' open and close. In fact the word snapdragon dates back to at least 1597, when Gerard suggested that women (rather than a serious male horticulturist such as himself) had invented the name. Gerard's primary name for the plant was Calves snout, referring to the seed capsules rather than the flower, though he added that 'in mine opinion it is more like the bones of a sheeps head that hath beene long in the water, the flesh consumed cleane away'. One can see why the name snapdragon caught on – though in Gerard's time

Snapdragon, *Antirrhinum majus*

the Spanish called it *cabeza del tenera* and the French *tête de veaux*, both meaning calves' head, and today it is known in France as *muffle* or 'muzzle'. Another old English name for the plant, Lyons snap, survives in the present-day Italian *bocca di leone*. *Linaria*, incidentally, gets its English name of toadflax from the notion that its flowers, which look like miniature snapdragons, resemble the mouths of toads or frogs.

Like the heart, the hand is generally used in botanical Latin to refer to leaf shape, and sometimes as little resembles the part of the body from which it takes its name. *Palma*, which is fairly obviously the Latin for hand, gives us *palmatus* for hand-shaped. It is no good, however, counting the lobes on a leaf and expecting them to number five, like the fingers on a human hand. The Japanese maple *Acer palmatum*, for example, always has an odd number of lobes, but these might number five, seven or even nine, and even the leaves of the emphatically named subspecies *A. palmatum* subsp. *palmatum* can have seven lobes. The leaves of the Chinese *Kirengeshoma palmata* I once grew barely resembled hands at all, and are more accurately described by the RHS *Index* as 'broadly ovate, palmatifid to unevenly incised'. Palmatifid means 'palmately cleft rather than lobed', but frankly there is little of the hand about these leaves – though they are certainly different from the leaves of *K. koreana*. It is possible that the common wallflower got its original botanical name from the hand. Now *Erysimum cheiri*, it is still sometimes called *Cheiranthus cheiri*, and both the genus and species name were once thought to derive from the Greek *kheir*, meaning hand, the same word that gives us the chiropractor, who uses hands to diagnose and treat problems in the joints and spine, and chiromancy, the reading of palms. The botanical etymology is thought to refer to the notion that these heavily scented flowers were often carried in the hand by those attending medieval festivals; but another suggestion is that the name derives from *kheyri*, the Arabic word for the plant meaning 'red-flowered', because although the flowers come in many tawny shades (and it was once known as the yellow stock gillyflower), dark red was particularly prized, as in the aptly named ancient cultivar 'Bloody Warrior'. The foxglove, meanwhile, goes to *digitus*, the Latin word for finger, for its botanical name, *Digitalis*. This was coined by the sixteenth-century botanist Leonhart Fuchs from the already existing German name of *Fingerhut*, meaning 'finger-hat', referring to the way children picked the flowers to slip on their fingers like little hoods.

The *Omphalodes* genus takes its name from the Greek word for navel, *omphalos*, referring to the shape of the seeds. The true Venus's navelwort is *Omphalodes linifolia*, a beautiful plant with white flowers like those of the forget-me-not and grey-green foliage, but the English name is freely applied to other species such as the blue-flowered *O. cappadocica*, of which the regrettably popular cultivar 'Starry Eyes' has white flowers with feathered markings in a strong blue. Easier to see than seeds for identification purposes are leaves, and those of the wall pennywort, *Umbilicus rupestris* (the genus name of which is the Latin equivalent of *omphalos*),

Common foxglove,
Digitalis purpurea

have the smooth surface of a child's belly with a distinctly navel-like depression at their centre. Wall pennywort was also once known as navelwort, this species even having an early botanical name of *Umbilicus veneris*, evoking Venus. Gerard gives a further English name of kidneywort, though this name has now been applied by those with a hazy sense of anatomy to *Anemone hepatica*, the species name of which in fact refers to the liver, the Greek word for that vital organ being *hepar* (gen. *hepatos*). This plant is also known, and by many more people, as hepatica *tout court*, or liverwort, but at the time of writing there is some disagreement in the rarefied world of botanical nomenclature as to whether *Hepatica* is a genus of its own or a species more truly belonging to the anemones. More to the point is why it was thought their leaves resemble the liver, an organ which (unlike the kidney) is difficult to define in terms of its shape. Livers do, however, have lobes, and the leaves of some hepaticas are lobed, notably the ones most commonly grown in gardens: *H. nobilis*, and more particularly (as the name tells us) *H. acutiloba*. The flowers are usually a vivid violet and the leaves are green and sometimes tinged with purple, which rather confounds the notion parroted in many dictionaries of botanical Latin that *hepaticus* means 'liver brown'.

Lungwort, *Pulmonaria officinalis*, has many other common names, including Jerusalem cowslip, although it isn't a primula and is a woodland plant so unlikely to be found growing anywhere near the Holy City. The flowers do in fact resemble those of the cowslip but are variable in colour from pink to blue. It is, however, the spotted leaves that give the plant its botanical name, from their perceived resemblance to a diseased lung, *pulmo* being the Latin name for that organ and familiar as the root of several anatomical and medical words. The doctrine of signatures led to lungwort being used by early physicians and apothecaries for pulmonary complaints, and even today a bitter-tasting tea made from the plant's dried leaves is peddled by herbalists who claim it has a beneficial effect on the respiratory system. To others the white spots on the leaves suggested the milk or tears of the Virgin, a notion that gave rise to a number of regional names such as Lady's Milk-Sile (or milk-strainer) and Lady Mary's Tears, and it may be this religious association that led to the plant being linked to the Holy Land, since it has also been called Bethlehem sage.

Leaves with the more easily identifiable shape of a kidney are common enough for the word reniform (from the Latin *renes*) to have entered botanical glossaries. The small kidney-shaped leaves of the creeping pink-flowered plant *Hypsela reniformis* are not as obviously reniform as the larger ones of the carnivorous Brazilian *Utricularia reniformis*, though confusingly the latter plant is a member of the bladderwort genus, so called because of the bladders in which the plant captures its prey, *utriculus* being the Latin word for a small bottle. The large leaves of *Ligularia reniformis* have earned it the name of the tractor-seat plant. This will bemuse younger gardeners, but older ones may recall the pierced kidney-shaped metal seats that were a not very comfortable feature of early tractors made by such firms as Bamford or Fordson.

Kidney vetch, a pretty yellow European wildflower, was once thought to 'prevayle much against the hoate pisse, the stranguary or difficultie to make water, and against the payne of the Reynes'. Like many such plants, however, its supposed 'vertues' went beyond the renal, and its botanical name, *Anthyllis vulneraria* (from the Latin *vulnus*, a wound), is reflected its reputation for more general healing. The eminently sensible Anne Pratt, a leading Victorian authority on wild plants, writes that it was Conrad Gessner in the sixteenth century who 'first ascribed to this plant its vulnerary properties, but saving that it is downy and soft as lint, these are not very apparent'. The genus name, coined by Dioscorides, indeed means 'downy flower'.

The name woundwort has been applied over the years to a number of plants, including the aquatic water soldier, *Stratiotes aloides*, which Gerard claimed 'staieth the bloud which commeth from the kidneies' and 'keepeth green wounds from being inflamed'. The name is now more usually used for *Stachys sylvatica*, which despite its specific woodland epithet is known as hedge woundwort. Another kind of all-heal is true valerian, the tall, pale pink *Valeriana officinalis*, not be confused with red valerian, *Centranthus ruber*. It takes its name from *valere*, the Latin verb meaning to be healthy, and its healing properties were so highly regarded, Gerard writes, that 'among the poore people of our Northerne parts . . . no broths, pottage, or physical meats are worth any thing' unless it was also served at table. Goldenrod is another plant once known as woundwort, and it takes its botanical name, *Solidago*, from the Latin *solidere*, meaning to make whole or to strengthen. In Gerard's time it was used as a diuretic and to dissolve kidney stones and was 'extolled aboue all other herbes for the stopping of bloud in sanguinolent vlcers and bleeding wounds'.

Cinquefoils were also considered a useful cure-all and their botanical name, *Potentilla*, is a diminutive of the Latin *potens*, meaning powerful. Gerard recorded that *Potentilla anserina*, or silverweed, can be used to cure 'bloudy flix, and all other flux of bloud in man or woman', mouth ulcers, and injuries incurred by 'falling from some high place', and that it 'hath many other good vertues, especially against the stone, inward wounds, and wounds of the priuie or secret parts'. Comfrey is now grown ornamentally and put to practical use as a highly effective, if noisome, manure, but it was once known as knitbone or

'Wall Cinkfoile', probably
Potentilla erecta

boneset because of its supposed efficacy in mending fractures. It takes its English name from *confervere*, a Latin word meaning to heal broken bones, while its botanical name, *Symphytum*, is derived from the Greek *syn* = together + *physis* = growth.

As well as the doctrine of signatures, some early physicians believed that the properties plants showed in their natural habitat suggested a possible medicinal use. For example, the supposed rock-rupturing qualities of saxifrage, embodied in its name, led doctors to imagine the plant would be effective in breaking up kidney stones. The use of plants to cure or alleviate assorted ailments resulted in the epithet *officinalis*, taken from *officina*, the Latin word for workshop, which later became associated with apothecaries' shops. It denotes plants that might be obtained from such establishments, or more broadly any plants that were used medicinally. The common sage, *Salvia officinalis*, is now chiefly used in cookery, but once served as a remedy for a wide range of ills, from urine retention to rheumatism. The genus name, from which we get the word salve, comes from the Latin *salvus*, meaning safe, well, or sound. Given that so many plants were used medicinally, specific epithets beyond *officinalis* were devised for plants with a more specialised application. Figwort, for example, was used to treat the lymphatic disease King's Evil or scrofula, 'or any other knots, kernels, bunches or wennes growing in the flesh whatsoever', including haemorrhoids, which were known as figs. The reason for this medical application is that the plant's roots have conspicuous nodes and its dark red flowers could be said to resemble bloody nodules, hence the botanical name *Scrophularia nodosa*. Rather more reliable, because they contain tannin and therefore 'have a drying and astringent quality', as Parkinson correctly wrote in his *Theatrum Botanicum*, were burnets. Parkinson added: 'they are availeable in all manner of fluxes of blood, or humours, to stench [staunch] bleeding inward or outward', and he recommended them particularly for 'the Blooddy-flix' or dysentery, and 'womens too abundant courses'. It was for this reason that burnets gained the genus name of *Sanguisorba*, from the Latin *sanguis* = blood + *sorbere* = to soak up (as in 'absorb'). The wild service tree's botanical name, *Sorbus torminalis*, is sometimes said to be related to colic, but *tormina* is a Latin word for dysentery. Although we had one on the Herefordshire farm where I grew up, the wild service tree is now rare in Britain, but at one time its berries were widely used to treat not only dysentery but other intestinal or stomach problems.

A favourite remedy for all kinds of ailments was purging, in which patients were given various herbs to clean out the

Great burnet, *Sanguisorba officinalis*

digestive system. There are fourteen entries for purgings in 'The Table of Vertues' printed at the back of Gerard's *Herball*, and it is clear that a large number of plants were used for this somewhat drastic cure-all. Catharsis is a word more usually used today in an emotional or psychological context, but the botanical epithet for plants used as purgatives, *catharticus*, comes from the same Greek root: *kathairein*, to cleanse, from *katharos*, pure. The purging buckthorn, *Rhamnus cathartica*, is the principal host plant for the common brimstone butterfly (which takes its scientific name, *Gonepteryx rhamni*, from this shrub), and is now often grown to attract them; but it was once so popular as a purgative that it was also known as 'Laxative Ram'. According to Gerard, the best use of the shrub's black berries was 'to breake them and boile them in a fat flesh broth without salt, and to give the broth to drink; for so they purge with lesser trouble and fewer gripings' than when administered in powdered form. He adds that the berries were also mixed with alum by painters to produce a colour called 'Sap greene'.

The opium poppy *Papaver somniferum* is grown in gardens for its glaucous foliage, and flowers that range from pale mauve to very dark purple, but elsewhere vast fields of the plant are grown for the production of the drug from which it takes its common name. This was the use to which these poppies were put by physicians in the past, often administered as diacodium, a syrup made from the seed heads. The poppy's specific epithet is taken from the Latin *somnus* (sleep) + *ferre* (to bring), and diacodium was widely used as a soporific and to alleviate pains of various sorts. Although this remedy was prescribed by leading doctors of their day, such as Sir Hans Sloane, it was so simple to make that recipes can be found in household recipe books such as the one someone in my family compiled in the late seventeenth or early eighteenth century. The first two pages of the book are

missing, but the fact that two recipes for diacodium appear on
the third page suggests how important this concoction was. All
that appears to have been required was a hundred or so poppy
heads cut into quarters, some spring water to boil them in, and
a pound of sugar to every pint of the resulting liquor.

There are records of Sloane administering diacodium in the
case of severe, convulsive coughs, but in less grave cases the
more usual remedy was coltsfoot, which takes its botanical
name *Tussilago farfara* from the Latin *tussis*, a cough, + *ago*, a
word relating to action. (The specific epithet is derived from *far-
reus*, meaning grain, referring to the mealy surface of the plant's
leaves.) It is the 'fresh leaves or iuyce or Syrup made thereof'
that Parkinson recommends to soothe 'an hot, dry cough', and
coltsfoot has remained an ingredient of cough syrups and loz-
enges for many centuries. Staying with throat ailments, there
is some uncertainty as to how self-heal, *Prunella vulgaris*, got its

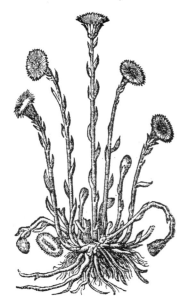

botanical name. In old German
herbals the name was sometimes spelled *Brunella*, which
may be derived from *braun*
(brown), perhaps referring to
the plant's stubby little flower-
heads, made up of clusters of
bright reddish brown calyces,
from which the purple flowers
emerge. However, the name
might equally come from what
Parkinson calls *die Bruen*, a
German term for an unidentified infection, the symptoms
of which were 'inflammation

Coltsfoot, *Tussilago farfara*   and swelling both in the mouth and throate, the tongue rough and rugged or black, and a fierce hot continuall ague thereon'. This disease has been variously identified as quinsy, an infection of the tonsils that is now comparatively rare but greatly troubled our ancestors, or diphtheria. Whatever the case, a decoction of self-heal, both drunk and used as a mouthwash, was the principal remedy. Also used to treat the throat was the nettle-leaved bellflower, *Campanula trachelium*, which takes its species name from *trakhelos*, the easily recognisable Greek word for both the neck and the throat. Gerard placed 'Throat-wort or Canterbury-Bells' in their own genus, *Trachelium*, separated out from *Campanula*, which he called 'Peach-bells and Steeple-bells'. The latter had no 'vertues', but *Trachelium* was recommended as a gargle for throat and mouth ailments as well as to make 'lotions or washing waters, to inject into the privy parts of man or woman'.

And it is with these privy parts that we end this brief tour of the body and its ills. It hardly needs explaining how the innocent-sounding butterfly pea got its botanical name of *Clitoria ternatea*, though anyone in doubt need only look at the flower. Linnaeus was merely following earlier botanists when he gave the plant this Latin binomial, since it had been first described seventy-five years before by a presumably startled German botanist working for the Dutch East India Company, who dubbed it *Flos clitoridis ternatinisibus*. Botanists in the Victorian era attempted to come up with less explicit, and sometimes scarcely relevant or simply ludicrous alternatives to the genus name: *Vexillaria* is taken from the Latin *vexillum*, a banner, while *Nauchea* appears to be related to the specific epithet *naucinus*, meaning modest. Anything less modest than the flower of the clitoria would be hard to imagine, although the flowers of some oriental orchids have also been known to alarm the gynaecologically squeamish. Orchids, however, derive their name from the male anatomy, the genus *Orchis* being a Latinised form of *orkhis*, the Greek word for testicle. Gerard listed several kinds of orchids with common names linking men or animals with 'stones' or 'cullions', the latter derived from the Latin *culleus*, a scrotum or testicle: 'Fooles stones', 'Goat stones', 'Fox stones'. The last category included 'Souldier's satyrion', otherwise the military orchid, *O. militaris*. This British native, now fabulously rare and ferociously protected, obsessed the writer and amateur botanist Jocelyn Brooke, who gave its name to the first volume of his *Orchid Trilogy* of autobiographical novels (1948–50). Satyrion was another old word for the orchid, derived from the Greek word *satyros* or satyr, those notoriously priapic woodland gods of mythology. The reason the plant acquired this name is evident from Gerard's description of the 'fat stalke full of sap or iuyce' that 'riseth up' from the 'stones'. Similarly, Geoffrey Grigson in his *Dictionary of English Plant Names* (1974) identifies 'the long purples' of Ophelia's garlands in *Hamlet*, to which 'liberal shepherds give a grosser name', as the early purple orchid, *Orchis mascula*, 'which has the shape of an erection above the two root-tubers' – and indeed the plant's specific epithet is taken from *masculus*, the Latin word for virile. Shakespeare's 'cold maids' called the plant 'dead men's fingers', and other orchids were reclassified as *Dactylorhiza*, from the Greek *dactylos* (finger) + *rhiza* (root), on the grounds that the tubers were more elongated than those that remained in the *Orchis* genus. Another group of orchids, including the bee orchid, have the genus name *Ophrys*, which is the Greek word for eyebrow and refers to the furred lips of the plants.

Although most gardeners select and grow plants for the initial visual impact they make in pots, beds and borders, other senses – smell, touch and taste – can also be an important consideration. Botanical names can both entice and warn the gardener, from *odoratus* (sweet-smelling) to *foetidus* (foul-smelling), from *mollis* (soft) to *pruriens* (itch-causing), and from *esculentus* (edible) to *venenosus* (highly poisonous). Some of these we could guess from English words derived from the same roots: *odoratus* clearly relates to odour, *foetidus* to foetid and *mollis* to emollient. *Pruriens* we might recognise from prurience, that unseemly interest in sexual matters encouraged by tabloid newspapers, and the botanical meaning takes us back to the Latin origins of the word: causing to itch.

Most poisonous plants are not botanically labelled as such, but it is clearly useful to be able to differentiate between the angelica that is candied for use as a cake decoration, *Angelica archangelica*, and its highly poisonous relative, *A. venenosa*. Epithets referring to edibility such as *esculentus* and the synonym *edulis* are more or less only used for fruit and vegetables, as is *deliciosus*. While it is quite clear why the kiwi fruit is *Actinidia deliciosa*, it will be less clear – in Britain at least – why the Swiss cheese plant is *Monstera deliciosa*. The common name refers to the plant's leaves which have holes in them vaguely reminiscent of Emmental cheese, and it is for these leaves that the plant is grown in Britain. In the tropics, however, the plant produces fruit which are poisonous until they ripen fully – a process that can take up to a year but is apparently worth waiting for, as the taste is likened to pineapple and banana.

Outside the kitchen garden, specific epithets relating to taste and toxicity are a good deal less useful now that we grow plants to look at or smell rather than consume for pharmacological reasons. In the days of the apothecaries and herbalists, however, it was handy to know whether the plants with which you were dosing your patients tasted pleasant or unpleasant, and therefore what other ingredients you might have to add to make them palatable. Unsurprisingly, honey seems to have had as important a place on the shelves as the various herbs used in pills and potions. It was believed that anything with a bitter taste would cleanse the body, which meant that, alongside rue, physicians favoured such plants as wormwood, *Artemisia absinthium*, which also had associations with sorrow (as in the biblical phrase 'wormwood and gall'), but was later used to flavour absinthe and other alcoholic drinks. The specific epithets for bitter are *amarus* and *amarellus*, and the yellow daisy *Helenium amarum* is known as bitterweed – though it is sufficiently toxic not to have been used in medicine. The autumn gentian, *Gentianella amarella*, however, was used as a cleansing tonic and was known as felwort, from the Latin *fel*, meaning gall, bile or bitterness. A garden-worthy British native with upturned lilac bell-shaped flowers, it was a welcome replacement for the related *Gentiana lutea*, which had to be imported and was therefore expensive. The gentian was named after King Gentius of Illyria (fl. 181–168 BC) by Dioscorides, in the mistaken belief that this monarch had first discovered the medical properties of the plant. In fact, there is a record (written on papyrus discovered with a mummy in Thebes) of gentian being used for medical purposes by the Egyptians more than a thousand years earlier. The use in England of *Gentianella amarella* and its fellow native *G. campestris* was advocated by the celebrated seventeenth-century herbalist Nicholas Culpeper, who claimed that 'the Reasons so frequently alledged, Why English Herbs should be fittest for English Bodies, hath been proved by the experience of divers Physitians'.

Another epithet meaning bitter-tasting is *acris*, familiar from the meadow buttercup *Ranunculus acris*, the 'Flore Pleno' cultivar of which, with its golden button flowers, is often grown in gardens. For anyone foolish enough to put it to the test, the taste of a buttercup is indeed acrid; in fact this much-loved plant, traditionally held under the chins of children to detect whether or not they like butter, is highly poisonous: ingesting it can cause ulceration of the mouth, followed by damage to the digestive system, convulsions and death. Plants that are sharp-tasting rather than bitter take the epithets *acetosus* or *acetosellus*, derived from *acetum*, the Latin word for vinegar, and found in the names of various types of sorrel: wood sorrel is *Oxalis acetosella*, sheep's sorrel *Rumex acetosella*, and the garden sorrel grown for cooking is *Rumex acetosa*. The equally delicious buckler's sorrel, *Rumex scutatus*, derives both its common and botanical names from the shape of its leaves, *scutum* being the Latin for a small shield that Roman soldiers often wore strapped to their forearms. It is both the garden-sorrel shape and acidic taste of the leaves of *Pelargonium acetosum* that give this plant its name. The Greek word for sour is *oxys*, which not only gives us the genus name of the wood sorrel but the epithet *oxycarpus*, meaning sour-fruited – not

that one is likely to try the fruit of the claret ash, *Fraxinus oxycarpa* 'Raywood', grown for its glowing autumn colour, or of the yellow rock-plant *Alyssum oxycarpum*. Similarly, although the roots of the tropical climber *Bomarea salsilla* are edible, and presumably taste as their name

'White Wood Sorrell', *Oxalis acetocella*

suggests (*salsus* is the Latin word for salty), gardeners are more likely to grow the plant for the reddish pink trumpet flowers, which also show that it is related to alstroemeria. The *Ribes* genus, which includes blackcurrants and gooseberries, derives its name from *ribas*, a Persian word meaning sour.

The botanical opposite of these salty, sour and bitter epithets is *dulcis*, the Latin word meaning sweet or pleasant. The sweet almond is *Prunus dulcis*, and the Japanese raisin tree is called *Hovenia dulcis* because not only its fruit but also its leaves are sweet to the taste. The most familiar garden plant with this epithet, however, is the sweet spurge, *Euphorbia dulcis*, particularly popular as the 'Chameleon' cultivar, which as this name suggests goes through several changes of colour during the season. It is conceivable that, compared with other spurges, the taste of this one is mild, but it is not something anyone sensible would put to the test. As Gerard notes of spurges: 'the iuice or milke is good to stop hollow teeth, being put into them warily, so that you touch neither the gums, nor any of the other teeth in the mouth'. The Latin word *dulcis* can also mean pleasant or charming, which may be how the spurge got its name, and *blandus* is more or less a synonym, both in Latin and as a specific epithet.

And while we are dealing with such nebulous concepts as charm and pleasantness, we should mention again the specific epithet *tristis*. When not referring to sombre or pallid colours, *tristis* can also indicate plants with a drooping and therefore 'mournful' habit, as in the cultivar name of the weeping silver birch, *Betula pendula* 'Tristis', or plants that are at their best at night. Although *Pelargonium triste* may have got its name from its rather dingily coloured flowers, these only smell at night and, acccording to the Rev. Samuel Gilbert's *The Florist's Vade Mecum* (1683), the plant is *triste* because 'it rejoiceth not by day'. The night-flowering jasmine *Nyctanthes arbor-tristis* is sometimes called the tree of sorrow both because of its weeping habit and because its flowers fall as soon as the sun is up. According to Hindu mythology, the plant is one of the five that grant human wishes, which is why it is used in religious rituals. The story of its origins is indeed sad. A princess called Parijata fell in love with the sun god, Surya, who told her that if she wanted to be with him she would have to renounce her kingdom, leaving behind her royal clothes and possessions. She agreed to this, but because Surya was the sun, his passion flamed and soon burned out, as is the nature of sexual desire, after which he returned to the sky. The princess died of a broken heart, and from the ashes of her funeral pyre *Nyctanthes arbor-tristis* grew, its branches bowed down with mourning. It flowers only at night, the blossom falling and dying as the sun's first rays light the earth, which is when it has the strongest fragrance. In one version of the story, a strong wind carried the princess's ashes into the air and wherever they fell a night-flowering jasmine grew, hence its widespread distribution throughout India.

Scent is a major consideration when we decide what to grow in our gardens, and there are several specific epithets that alert us to this property. Among the general ones for a pleasant smell are *aromaticus, fragrans, odorus, odoratus* and *odoratissimus*, all of which will be familiar from similarly derived English words. Among the spices, the clove is *Syzygium aromaticum* and the nutmeg *Myristica fragrans*, both of them members of the reliably scented *Myrtaceae* or myrtle family. In the case of the fragrant sumach, *Rhus aromatica*, it is the leaves and stems rather than the flower buds and fruits that have a pleasant, lemony smell. Several species of clematis have flowers with a good scent, including the early-flowering evergreen *Clematis armandii*, the scented virgin's bower, *C. flammula*, and the cowslip-scented *C. rehderiana*; but only two refer to their fragrance in their names, the summer-flowering and sweetly scented purple-starred *C. × aromatica* and the spring-flowering purple-belled *C. koreana* var. *fragrans*, which is reputed to smell of lime and sandalwood. Among other shrubs, both the sweet osmanthus, *Osmanthus fragrans* (the genus name of which combines the Greek *osme*, odour, with *anthos*), and the sweet viburnum, *Viburnum fragrans* (syn. *V. farreri*), are valued for their strongly scented flowers, but perhaps the best loved is *Daphne odora*. Few would grow daphnes for their appearance, which is frankly dreary even in the variegated forms; but in late winter the air can be suffused with the sweet-sharp fragrance of their otherwise inconspicuous flowers.

Plants with English names beginning 'sweet' often take the specific epithet *odoratus*, including sweet cicely (*Myrrhis odorata*), sweet woodruff (*Galium odoratum*), the sweet violet (*Viola odorata*), and of course the sweet pea (*Lathyrus odoratus*), the scent of which is for some the very essence of an English summer. Although not as strongly perfumed as many other roses, the tea rose from China was widely grown in the early nineteenth century and was designated *Rosa × odorata*. Hugely popular in the Victorian period among the poorer classes both because of its sweet smell and because it was sold cheap in the streets, was mignonette, *Reseda odorata*, introduced to Britain in the 1740s. One of the street vendors that Henry Mayhew interviewed for his pioneering sociological study *London Labour and the London Poor* (1851) claimed to be able to sell 600 penny pots of mignonette in a single day. 'Mignonette's everybody's money,' he said, and the flowers' scent no doubt helped to mask the all too pervasive everyday smells of slum-dwelling.

Scented-leaved pelargoniums are now familiar in many species, hybrids and cultivars, but they were a comparatively late introduction, unknown to either Gerard or Parkinson. The pelargonium arrived properly in Britain in the eighteenth century, but there is a story that John Tradescant the Elder may have brought back a specimen of *Pelargonium triste* from Algiers as early as 1620. This plant, like most of the scented-leaved pelargoniums, is in fact a native of South Africa, so if the story is true it must be assumed that someone had already imported the plant to North Africa. Another tale is that *P. triste* had been brought to Britain in the 1630s by sailors returning from India, for which reason it was mistakenly thought to grow there and so was dubbed *Geranium indicum noctu odoratum* (*noctu* being Latin for 'by night'). In fact the sailors would have stopped at the Cape on their voyage home, and this would have been where they found the plant, since it is not native to India. The species with a name meaning very scented, *P. odorantissimum*, has a very distinct smell of apples; crossed with *P. exstipulatum* it produced the hybrid *P. × fragrans*, which smells of eucalyptus but has been the parent of numerous cultivars smelling of such spices as cinnamon and nutmeg. The fragrance most gardeners associate with scented-leaved pelargoniums is rose or Turkish delight, and these plants are sometimes known as 'rose geraniums', a botanically incorrect term also used for perfumes and soaps. The source of such cultivars as 'Attar of Roses' is *P. graveolens*, the specific epithet of which is taken from the Latin *gravis* = strong + *olere* = to smell. Although this pelargonium's scent, though undeniably strong, is pleasant, *graveolens* can also mean rank, as in common rue, *Ruta graveolens*. The epithet *suaveolens* is less ambiguous, *suavis* being the Latin for sweet or delightful, and there is indeed a *Pelargonium suaveolens* (though the name is disputed), most familiar in the cultivar 'Orange Fizz', the leaves of which smell like the kind of orangeade that fizzes with E numbers. Other plants that take this epithet include the Australian tobacco *Nicotiana suaveolens*, the white angel's trumpet *Brugmansia suaveolens*, and the dwarf iris *Iris suaveolens*, all of which have notably sweet-scented flowers.

It would seem that you can now find scented-leaved pelargoniums with almost any smell you like, from peppermint and pine, through assorted fruits (including strawberry, lemon and gooseberry) to almond, anise and 'roasted nuts'. Some of these smells have their own specific epithets when used in the names of other kinds of plant. The star anise is *Illicium anisatum*, taking its genus name from the Latin verb *illicere*, meaning to allure or entice – in this case by its smell. Similarly scented plants include

the ringwood tree, *Syzygium anisatum*, and the giant purple-flowered hyssop *Agastache anisata*. The latter is now more usually called *A. foeniculum*, taking its name from fennel (*Foeniculum vulgare*), which has a less strong but similar odour to aniseed. A saffron-like scent gives the *Crocosmia* genus its name, from the Greek *krokos* + *osme* = a smell. Balsam is another distinctive smell, named after the fragrant resin exuded by several different trees and shrubs, but traditionally by the Arabian balsam, *Commiphora gileadensis* (the

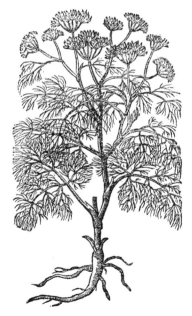

Common fennel,
Foeniculum vulgare

genus name of which is the Greek for 'resin-bearing'). One
of the most resonant as well as resinous smells of my childhood was the one that filled the garden after summer rain had
released it from the rustling leaves of the balsam poplar, *Populus
balsamifera*. This is not really a suitable tree for gardens, since
it tends to throw up saplings from its extensive and sometimes
structure-damaging root runs, but the scent is irresistible. The
Hudson pine, *Abies balsamea*, is similarly fragrant, but similarly
problematic, since it can reach a height of over twenty metres;
there is, however, a dwarf form, *A. balsamea* 'Nana', the leaves
of which can easily be reached and crushed to release the scent.
Perhaps the best solution for those who crave the smell is to
grow the tender perennial *Cedronella canariensis*, sometimes
called balm-of-Gilead, which not only has the balsam scent but
short spikes of attractive pink
labiate flowers. Although it has
its own distinctive smell, peppermint is sometimes called
Mentha balsamea; but it is
more usually and sensibly *M.* ×
piperata, from the Latin *piper*,
meaning pepper and referring
to the leaves' strong taste.

Carnation or clove pink,
Dianthus caryophyllus

Linnaeus called the clove
tree *Caryophyllus aromatica*,
and his genus name (from the
Greek *karyon* = nut + *phyllon*
= leaf) survives as a specific
epithet for plants that smell of
cloves, notably the carnation,
Dianthus caryophyllus. Gerard
gave both carnations and pinks

the genus name of *Caryophyllus*, and these were among several plants with strongly scented blooms (including stocks, dame's rocket and wallflowers) that were commonly called gillyflowers, a corruption of the Old French word for a clove, *gilofre*. The present-day French word is *girofle*, or more fully *clou de girofle*, *clou* (meaning nail) giving us the English word 'clove'; carnations are also known as clove pinks. The carnation's specific epithet has been adapted for other plants with similarly spice-scented flowers, such as *Freesia caryophyllacea* and *Alstroemeria caryophyllacea*. Lemon-scented plants take the epithet *citriodorus*: lemon thyme is *Thymus citriodorus*, lemon basil *Ocimum* × *citriodorum*, and the lemon-scented gum *Eucalyptus citriodora*. *Pelargonium* 'Citriodorum' is a cultivar with citrus-scented leaves; but the equally fragrant lemon verbena drops the 'i' and is *Aloysia citrodora*. In the British climate lemon often goes with honey, for which the Latin word is *mel* (gen. *mellis*), giving us *Euphorbia mellifera* for the honey spurge, *Eucalyptus melliodora* for yellow box, and *Iris mellita* as an old name for *I. suaveolens* mentioned above. The tender architectural shrub *Melianthus major* takes its genus name from the same source and means 'honey-flower', because the dramatic dark russet flower-spikes that rise (if you are lucky here in Britain) from its large grey-green serrated leaves are charged with nectar.

The word musk has a similar derivation to 'orchid', originating in the Sanskrit word *muska*, meaning scrotum. This is because the gland on the male musk-deer's abdomen that produces a substance used in the manufacture of perfume resembles a testicular pouch. The species name for plants with a musk-like scent is *moschata*, as in the musk mallow, *Malva moschata*. The rambling musk rose, *Rosa moschata*, was introduced into cultivation very early, and it is now unclear exactly where it came from, though it may have been the Himalayas. It was known in Britain at least by the mid-sixteenth century, was grown in several varieties by Gerard, and is referred to by Titania in *A Midsummer Night's Dream*, when she dispatches her fairies 'to kill cankers in the musk-rose buds'. Though not quite as long in British cultivation, *Narcissus moschatus* has been grown here since the early seventeenth century and is known as the swan-necked daffodil because of the way its scented white flowers droop gracefully from their stalks.

So much for the perfumes with which we like to fill our gardens. There are, of course, beautiful plants that have little or no scent, some of them odd-man-out species of genera normally noted for their fragrance. Some of the witch hazels with reddish flowers, such as *Hamamelis* × *intermedia* 'Jelena', may make a nice change from the more usual yellow-flowered ones, but they often lack scent. (The genus name of *Hamamelis*, incidentally, has nothing to do with honey but – as sometimes happens – is merely the Greek name for some other plant altogether, mistakenly transferred.) Shrubs of the *Mahonia* genus, which can look about as cheering as a Victorian funeral for much of the year, are nevertheless valued because their racemes or clusters of yellow flowers appear very early. Most of these have a strong scent reminiscent of lily of the valley, so it seems rather pointless to grow those that don't, and the same might be said of *Philadelphus*. In the case of this genus, however, the gardener is warned, and the scentless mock-orange is *Philadelphus inodorus*. The Greek equivalent of this epithet is *anosmus*, and there is a scentless orchid called *Dendrobium anosmum*. However, as its synonym, *D. superbum*, indicates, its flowers more than compensate for their lack of scent with their beauty.

Then there are plants we grow *in spite of* their smell. Graham

Stuart Thomas (1909–2003), the long-serving Gardens Advisor to the National Trust, was reported to have 'an acute and discerning sense of smell' and once famously likened *Salvia turkestanica* to 'barmaid's armpits', which may or may not seem pleasurable depending on your predilections. Like colour, smells can strike people in different ways. I was once visiting the Royal Mausoleum at Frogmore when lilac was in bloom. As Ivor Novello recognised in his hugely popular song 'We'll gather lilacs', this is a plant normally associated with young love and springtime; but perhaps because she was very old, or perhaps prompted by her funerary surrounding, my companion buried her nose in the blossom and said: 'The very smell of decay.' Others find the scent of lilies either too overpowering or too reminiscent of bedecked coffins sliding towards the furnace, and while this may be a minority opinion, few of us would want to get too near *Lilium pyrenaicum*, despite its tempting little spotted, yellow turk's-cap flowers. The specific epithet gives nothing away, merely informing us that the lily originates in the Pyrenees, but even the customary warning epithet *foetidum* (from the Latin *fetere*, to stink) would hardly do justice to the post-coital reek of this plant. On the other hand, I often wondered how *Helleborus foetidus*, the stinking hellebore, got its name. Dutifully bringing the handsome russet-edged green flowers to my nostrils and inhaling deeply, I could detect nothing. Then someone told me to rub the equally handsome dark green leaves between my fingers, and this does produce a smell that although not wholly repellent is not one you would actively seek out. It is also by crushing the leaves of *Viburnum foetidum* that you release their odour; according to W. J. Bean in his *Trees and Shrubs Hardy in the British Isles* (1914), this smell can linger in herbariums centuries after the specimens were gathered and mounted.

Not that such strong smells are always unwelcome. The leaves of *Eryngium foetidum*, for example, are widely used in South Asian cookery, and while not in the same nostril-assaulting league as *nam pla*, the nevertheless delicious sauce made from long-fermented fish, the plant still lives up to its name.

No one would want skunk cabbage, *Symplocarpus foetidus*, anywhere near a kitchen. It gives fair warning in both its common and botanical names, and has the sinister maroon and cream-left-too-long-in-the-fridge colouring of several insectivorous plants. Like them, the smell it emits has been likened to rotting meat – though others detect a waft of rotting turnips, which may be unfamiliar to city dwellers but really does knock you back if you ever come across them in a farmyard root-clamp. The exclamation of disgust you might make in such circumstances could be phonetically rendered in the epithet *phu*, from a Greek word *phou* meaning evil-smelling. The impact in an early spring border of the golden spikenard, *Valeriana phu* 'Aurea', is chiefly visual rather than olfactory, and the smell of the flowers has been described as like honey. The name, however, probably refers to the plant's roots. Spikenard is a very old and costly perfume, the one used by Mary Magdalene to anoint Jesus's feet in the Bible, and by the grieving Achilles while preparing Patroclus's corpse for the funeral pyre in the *Iliad*. True spikenard, however, comes from the rhizomes of another plant altogether, though one that is also a member of the valerian family, the Himalayan *Nardostachys jatamansi*. That said, it is not clear whether this was the spikenard of history, since the related *Valeriana tuberosa* and the unrelated *Lavandula stoechas* have also been suggested as possible candidates. Some people find that the roots of common valerian, *V. officinalis*, have a strong and unpleasant odour. Though I grow this plant and – since it seeds freely – am often digging up unwanted offspring, I cannot say I have noticed any particular smell from the roots, unpleasant or otherwise; but I have noticed that neighbouring cats sometimes scratch away at the base of these plants to expose the roots and then rub their faces against them, and this appears to affect them in the same way that nepeta does.

Although, like cats, humans sometimes rub or stroke plants to release compounds that attract us by their smell, many of the species names relating to touch act more as a warning than an enticement. Sometimes the botanical name is rather less misleading than the common one. The velvet bean, *Mucuna pruriens*, for example, is a very vigorous climber native to Africa and the Asian tropics. Although its panicles of dark purple flowers are highly decorative, you are unlikely to find it in a British garden – and this is perhaps as well. The flower calyces act as a serious irritant, as do the hairy seedpods that give it its common name. The appearance of the pods is indeed temptingly velvety, but the effects of the hairs on the human skin is so severe and long lasting, and provokes in its victims such a frenzy of scratching, that the plant is also, and more usefully, known as the mad bean.

Few garden plants are as unpleasant as this, but there are a number that can cause irritation if touched. The sap of the *Euphorbia* tribe has already been mentioned, but worse still is garden rue, which contains a chemical that can cause phytophotodermatitis, which is to say that skin coming into contact with it can become highly sensitive to sunlight, resulting in severe chemical burns. Gardeners, therefore, might well rue the day that they didn't wear gloves when cutting back the plant – and although the plant and that feeling of regret have a different etymology, the two are closely related in history. Despite its possible effect on the skin, rue was widely used in medicine from earliest times, recommended by both Pliny and Dioscorides, and the plant's 'vertues' take up a whole page of Gerard's *Herball*. It is the bitterness of the plant's leaves that led to it being given the English name of rue, and although the origins of the botanical name of *Ruta* are disputed, it possibly comes from the Greek word *rhyte*, meaning unpleasant, which could also be a reference to its smell, confirmed, as we have seen, by the species name *graveolens*. Borage can also cause skin irritation, and although the etymology of its botanical name *Borago* is uncertain (Linnaeus thought it a corruption of *corago*, heart + action, because it was used to calm palpitations), it almost certainly refers to the hairiness of the plant's leaves. Possible derivations are from a medieval Latin word, *burra*, meaning a hairy garment – such as a 'hair shirt' – or from the Arabic *abū ḥurāš* meaning 'father of roughness'. Similarly hairy-leaved is the comfrey *Symphytum asperum*, from the Latin *asper*, meaning rough – though the word can also mean critical or rude, hence the casting of aspersions. A good deal rougher is the specific epithet *asperrimus*, applied (as *asperrimum*) to a white-flowered Australian heliotrope and a pink-flowered echium, both of which tend to be sold with warnings about their propensity for causing skin irritation. The genus of *Asperula*, of which the blue woodruff *A. orientalis* is a pretty bee-attracting garden annual, takes its name from the same Latin source but refers not to the plants' leaves but to their stems. Also meaning rough is the epithet *scaber* (as in 'scabrous'), though quite why it is used for the Chilean glory-vine, *Eccremocarpus scaber* is unclear – if ever a plant should be *scandens*, this is it, but the epithet perhaps refers to its wiry stems.

Taken directly from the Latin, the specific epithet *hispidus* tends to mean bristly, whereas *hirsutus* is merely hairy. The stems of the pink-flowered *Robinia hispida* have crimson bristles that might lure the unwary pruner into a false sense of security because they make it less easy to see the sharp thorns characteristic of the genus and particularly vicious in *R. pseudoacacia* and *R. × ambigua*. It is the bristly fruits of the epaulette tree that earn it the name *Pterostyrax hispida*, although its leaves also have distinct hackles along their edges. Among border or rockery perennials, and bristling very gently, is *Cotula hispida*, with bright yellow pompom flowers above mats of silvery foliage. Hirsute plants include *Acanthus hirsutus*, a dwarf species with flower-spikes covered in dense grey hairs, from which unexpectedly yellow flowers emerge; the shrub *Indigofera hirsuta*, which owes its name to the rust-coloured hairs on its stems; and the canary clover *Dorycnium hirsutum*, which has flowers tinged a pale pink and beautifully backlit by the silver hairs that cover the rest of the plant.

Other specific epithets denoting varying degrees of hairiness include *hirtus, capillaris, comatus, comosus, pilosus* and *villosus*. *Hirtus* is merely a variant of *hirsutus*, referring in the case of black-eyed Susan, *Rudbeckia hirta*, to the hairs of its leaves and stem, while *Polygonatum hirtum* is differentiated from other species of Solomon's seal by the pubescent (a botanical term meaning downy) undersides of its leaves. All the other epithets are derived from different Latin words for hair or hairiness. The Latin for a single hair is *pilus*, although botanically *pilosus* generally denotes plants with a great many soft hairs. The small teasel, *Dipsacus pilosus*, has flower-heads with woolly spines quite unlike the common teasel, *D. fullonum*, and the leaves, stems and buds of the apricot-flowered poppy *Papaver pilosum* have so many short soft hairs that they look felted. Similarly, *Heuchera pilosissima* is differentiated from other heucheras by the downy white hairs covering both its leaves and its flowers. *Capillus* is another Latin word

for a hair, from which we get
the epithet *capillaris*, easily
guessable from capillaries,
the thread-like blood-vessels
we all have. *Artemisia capillaris* gets its name because its
fragrant grey-green foliage
is even more finely divided
than that of other worm-
woods, and the same applies
to *Eupatorium capillifolium*,
sometimes called dog-fennel
and looking to my unappre-
ciative eye like a pondweed
that has become stranded on
dry land. The epithet *capillaris*
is also used for very fine spe-

Common teasel, *Dipsacus fullonum*

cies of such grasses as *Muhlenbergia* and *Agrostis*, while both
the common and botanical names of the innocent-sounding
maidenhair fern, *Adiantum capillus-veneris*, refer not to young
women and the goddess of love, but to the hair covering the
maidenhead or *mons Veneris* – as becomes clear if you look at
the wedgelike shape of the plant's individual leaflets.

Both *comosus* and *comatus* take their name from *comere*, a
Latin verb meaning to dress the hair. The deep violet sterile
flowers of the tassel grape hyacinth, *Muscari comosum*, sprout
from the top of the flower bud in a tuft of the kind people sport
when they tie up dreadlocks on top of their heads, though
the plant is mostly grown as the dragged-through-a-hedge-
backwards 'Plumosum' cultivar (*plumosus* being the Latin for
feathered). *Stipa comata*, sometimes known as the needle or
needle-and-thread grass, doesn't appear any more kempt, but

has long and very narrow inflorescences that look particularly attractive when a breeze blows through them. Another popular garden grass is *Pennisetum villosum*, sometimes known as feathertop grass because of its soft and hairy flower-heads. *Villosus* is the Latin word for shaggy, though botanically it is often used merely as another synonym for hairy, as in apple mint, *Mentha* × *villosa*, or the delicate, purple-flowered alpine *Soldanella villosa*, which has hairs on the stems and the undersides of its small lilypad leaves. The rough-sounding *Hydrangea aspera* is softened in its purple-flowered Villosa group, so called because their large leaves are like velvet to the touch.

From velvet leaves to plants whose varying degrees of prickliness have earned them warning specific epithets. One given in some dictionaries is *armiger*, a Latin word meaning armed, and therefore spiny or prickly, but this seems to be mainly restricted to fierce-looking ants and other variously bristling insects. More reliable is *horridus*, which does not simply mean generically unpleasant, but is from the Latin *horrere*, to bristle. The etymology of the English words horrid, horrible and horror is of course the same, since in its purest form horror was traditionally thought to make the hair stand on end, the technical term for which is horripilation (from *pilus* = hair; see above). Botanically, *horridus* means very prickly and is the species name of the cactus-like African milk barrel, *Euphorbia horrida*, and the pineapple sea holly, *Eryngium horridum*, which in addition to its sharp leaves has spiky-looking, pale green spherical flower-heads. I once visited a seaside garden full of succulents, including agaves, one of which had such sharp leaf tips that the owner had speared corks on them so that his dogs would not injure themselves. I can't now recall whether this was *Agave horrida*, but this species is one of those with spines along the leaf edge as well as at the leaf tip.

Plants with similar spines or thorns are often given the epithet *spinosus*, the Latin word for prickly (from *spina*, a thorn
or spine) – and, as in English, used of both plants and people.
The name of *Acanthus spinosus* will sound tautologous to classicists because the genus name is derived from *akantha*, the
Greek word for thorn, also familiar from *Pyracantha*, which
means fire-thorn, the prefix referring to the shrub's red berries. The specific epithet, however, distinguishes this species
of acanthus from *A. mollis*, the leaves of which are rounded
and without spines, *mollis* being the Latin for soft. (As with
spinosus, *mollis* could be applied to people, in this case those
who were considered weak or effeminate, hence 'mollycoddle'.)
The blackthorn, which produces pure white blossom in the
spring and purple-black sloes in the autumn, is *Prunus spinosus*
because of the ferocious thorns that make it such good stock-
proof hedging. Until its tubular two-tone red-and-orange
flowers appear, *Desfontainia spinosa* might easily be mistaken
for a holly from the shape and prickliness of its leaves, while
the species epithet of *Eryngium spinalba* adds colour and
refers to the ruff of prickly whitish bracts surrounding this sea
holly's flower-heads. Similarly the species name of the purple-
flowered alpine *Androsace spinulifera* indicates that the plant
carries small spines, in this case very small and not very sharp
ones at the leaf tips. And no one is going to prick themselves on
the 'spines' that give the spider plant, *Cleome spinosa*, its name,
since these are in fact extremely long anthers, which protrude
from the flowers to wave around like insect antennae.

It seems appropriate to end this chapter with a tailpiece,
belonging perhaps to animals rather than humans. The specific
epithet *caudatus* comes from the Latin word for a tail, *cauda*,
and generally distinguishes plants that have long, trailing
leaves or inflorescences. The trailing maidenhair fern, for

example, is *Adiantum caudatum* because when planted in a pot its long pinnate leaves hang down like tails over a gate. The long tassels of crimson flowers that characterise love-lies-bleeding may have suggested the gory ending of some traditional ballad to whoever gave it this name, but Linnaeus saw instead woolly red tails, and so named it *Amaranthus caudatus*.

'The body is the instrument of our hold upon the world,' Simone de Beauvoir observed, and it seems only natural that we should turn to it when naming the plants with which we have had so long and complex a relationship. No wonder that from head to tail, via many places in between, botanists throughout the centuries have looked to their own bodies, their senses and their ailments to put that world in some kind of order.

Word Lists

PARTS OF THE BODY

Antirrhinum = like a nose or snout

auriculatus = ear-shaped, earlobe-shaped

auritus = with long or large ears

capitatus = with dense flower-heads

caudatus = with a tail

cephalus = head

Clitoria = clitoris

hepatica = relating to the liver

macrocephalus = large-headed

microcephalus = small-headed

Nasturtium = nose-twisting, pungent

Omphalodes = navel

Ophrys = eyebrow

Orchis = testicle

palmatus = hand-shaped

Pulmonaria = relating to lungs

reniformis = kidney-shaped

testiculatus = testicle-shaped

trachelius = relating to the throat

vesicarius = relating to the bladder

MEDICAL

catharticus = purging

dysentericus = dysentery-curing

emeticus = emetic

febrifugus = fever-dispelling

medicus = used medicinally

officinalis = used medicinally

ptarmicus = causing to sneeze

somniferus = sleep-inducing

torminalis = dysentery-curing

trachelius = throat-treating

vomitorius = emetic

vulnerarius = wound-healing

TASTE

acetosus, acetosellus = sour
acris = bitter
amarus, amarellus = bitter
deliciosus = delicious
dulcis = sweet or pleasant
esculentus, edulis = edible

oxycarpus = sour-fruited
piperatus = peppery
saccharus = sugary
salsillus = salty
venenosus = poisonous

SMELL

agathosmus = smelling good
anisatus = anise-scented
anosmus = scentless
aromaticus = scented
balsameus, balsamiferus =
    balsam-scented
caryophyllus = clove-scented
citriodorus, citrodorus =
    lemon-scented
foetidus = stinking, foetid

fragrans = fragrant
graveolens = strongly scented
inodorus = unscented
melliferus, melliodorus, mellitus
    = honey-scented
moschatus – musk-scented
odoratus, odoratissimus, odorus
    = scented
phu = rotten-smelling
suaveolens = sweet-scented

TOUCH

aculeatus = prickly
armiger = spiny or prickly
asperus, asperrimus = rough
bombycinus = silky
capillaris = with long, fine
    hairs
capillifolius = with hair-like
    leaves

comatus, comosus = hairy,
    tufted
glutinosus = sticky
hirsutus, hirtus = hairy
hispidus = bristly
horridus = very prickly
pilosissimus = very shaggy
pilosus = with many soft hairs

pruriens = itch-causing

scaber = rough

spinosus = prickly

villosus = softly hairy, shaggy

MISCELLANEOUS

blandus = charming,
   pleasing

tristis = weeping,
   night-flowering

BESTIARY

From the time of Adam in the Garden of Eden, the naming
of animals and the naming of plants have often gone hand
in hand. Many early authors wrote books on natural history
which encompassed both the plant and the animal kingdom,
and we have seen how Theophrastus's *Historia plantarum* and
Aristotle's *Historia animalium* were translated into Latin at
the same time and by the same hand. As new plants were discovered, botanists often looked to animals in order to provide
names. In some cases, the colour or markings of leaf or flower,
or even the shape of a seed, suggested a creature. It was the spots
on the orange turk's-cap lily *Lilium pardalinum*, for example,
that prompted a specific epithet derived from *pardos*, the Greek
word for leopard, and the same notion led to the Australian
spotted hyacinth orchid being named *Dipodium pardalinum*
and the British spotted marsh orchid *Dactylorhiza pardalina*.
It was believed in the medieval period that the leopard was the
result of a lioness mating with a ferocious spotted creature called
a pard that was able to kill its prey in one leap. The pard in fact
exists only in medieval Bestiaries, though is referred to in both
Shakespeare's *As You Like It* and Keats's 'Ode to a Nightingale':

> *Away! away! for I will fly to thee,*
> *Not charioted by Bacchus and his pards,*
> *But on the viewless wings of Poesy . . .*

Keats is almost certainly using the word as a poetic archaism for leopard, since these are the creatures depicted drawing the god's chariot in such paintings as *Bacchus and Ariadne* by Titian, an artist the poet admired. If you weren't a god and so able to tame and harness these creatures, it was as well to have something to protect you from them, and the specific epithet of the yellow daisy popularly known as Leopard's bane, *Doronicum pardalianches*, means 'leopard-strangling'. A raw-meat bait laced with the plant apparently attracted and fatally poisoned wild animals that threatened shepherds and their flocks. The belief that some plants protected people from certain beasts was another way in which their botanical names came about.

In Europe wolves, rather than leopards, were the main threat to both livestock and humans. Gerard called some species of *Doronicum* Wolf's bane, an English name more usually associated with aconite, although the *Doronicum* and *Aconitum* genera are no more related than the animals they supposedly ward off or kill. According to Theophrastus, aconite derived its name from Akonia, a village of the Mariandynoi tribe, where the plant grew in abundance. This seems a rather obscure derivation, but one suggestion is that the now vanished village was in the Acheron region of north-west Greece, and the Acheron was one of the five rivers of the Underworld in Greek mythology. Another claim is that Akonia was in Anatolia, near a cave that supposedly marked the entrance to Hades. The sinister reputation of the plant as a source of quick transport to the realm of the dead would fit with either of these geographical associations. Although Theophrastus describes his aconite as 'a low-growing herb', it has been identified as the yellow-flowered *Aconitum anthora*, which, although not as tall as other species of monkshood, can grow to almost a metre in height. The specific epithet refers to the belief that the root of the plant would counteract the effects of thora, *Aconitum pardalianches* – which may be the same plant as the leopard-strangling doronicum. All aconites are highly toxic to humans as well as to animals. The species name of another yellow-flowered aconite, *A. lycoctonum*, means wolf-killing, and its very pale yellow subspecies *vulparia* (from the Latin *vulpes* = fox) is known as foxbane. Similarly, despite its specific epithet, the bright lime-green lichen *Letharia vulpina* is known as wolf lichen. The plant owes its colour to the vulpinic acid it contains, and its common and botanical names to the fact that is it was used to kill both wolves and foxes.

Not toxic, but something of a puzzle to early gardeners was the wolf peach, otherwise the tomato. According to Parkinson, the Persian scholar Avicenna (Ibn Sīnā, c.980–1037) condemned tomatoes, 'saying that those that are old are very noisome and hurtfull, although the fresh ones be better'. There were two known varieties in Parkinson's time, *Mala insana* or Madde Apples and *Poma amoris* or Apples of Love, both of which Gerard

Tomato, *Solanum lycopersicum* (syn. *Lycopersicon lycopersicum*)

grew, though more for their novelty than for any perceived 'vertues'. Gerard reports that 'in Egypt and Barbary they use to eat the fruit of *Mala insana* boiled or rosted under ashes, with oile, vineger, & pepper, as people use to eat Mushrooms. But I rather wish English men to content themselves with the meat and sauce of our owne Countrey, than with fruit and sauce eaten with such peril: for doubtless these apples have mischievous qualities, the use whereof is utterly to be forsaken.' Parkinson reassured his readers that mad apples did not in fact drive people insane, but that love apples might well incite 'Venery', as apparently evidenced by the relish with which Italians and the natives of other hot countries consumed them. Such is the popularity of tomatoes in Britain today, both to grow and to eat, that it is a surprise to find that even as late as 1724 the Scottish gardener and author James Gordon thought the 'apple of love' not worth growing, and certainly 'should by no means be planted on the sides of pleasure-walks, it having a most disagreeable smell'. Most people now would find the smell of growing tomatoes attractive; indeed the American perfumier Christopher Brosius said that it was while crawling among tomato vines as a child that he 'first discovered the pleasure of Scent', and he went on to manufacture and market a perfume that captured the moment: 'The shining green scent of tomato vines growing in the fresh earth of a country garden'. For Gordon, however, the tomato is merely a troublesome horticultural novelty: 'It is a very rampant grower, so is apt to over-run any plants that stand near it, and on the first appearance of frost makes a very dunghill of the place where it is planted; for which reasons, and the flowers also being of no beauty, it is scarcely worth the propagating above once.' There is no suggestion here that the fruit might be edible. By now the plant had gained its wolfish botanical name *Lycopersicon*, from *lykos* = wolf + *persikon* = peach, which appears to have been coined by Tournefort in 1694 in his three-volume *Éléments de Botanique*. Quite why is unclear, although it may relate to the fact that in German folklore members of the *Solanaceae*, which alongside the tomato include mandrake, henbane and deadly nightshade, were employed in witchcraft, and in particular to induce lycanthropy whereby men became werewolves. This may also be the reason for the name mad apple. Linnaeus retained the notion of the wolf peach in the specific epithet he used when he renamed the tomato *Solanum lycopersicum*.

Other reasons for using the wolf to name plants were more metaphorical. The specific epithet of the common hop, *Humulus lupulus* (meaning 'little wolf'), indicates that the plant is as rapacious in its growth as the wolf traditionally is in its appetite. The lupin is also generally considered to owe its name to the wolf, as seems even more clear from the American spelling, lupine, which is also the English adjective for wolf-like. The *Lupinus* genus, cultivated at first for animal fodder, was considered wolf-like in its ravaging of fertility, supposedly leaching all the goodness out of soil, and the plant was known in the Netherlands and Sweden as the wolf bean. (In the medieval period, prostitutes were sometimes called wolves 'because they devastate the possessions of their lovers'.) Alice Coats, however, suggested that this false reputation was due to a confusion between the Latin *lupus* and what she considered the true derivation from the Greek *lype*, meaning grief. In his *Georgics* Virgil refers to '*tristisque lupini*', usually translated as 'and the bitter [or sour or sad] lupin'. Coats ingeniously suggests that the adjective is explained by the fact that 'the seeds, unless boiled in several waters, are so bitter that the face of anyone eating them is drawn into an involuntary grimace of woe'. In other words, Virgil was making a play on the two words and their sad connotations. So perhaps the lupin does not really belong in a bestiary after all.

Many botanical dictionaries give *vulparius* as meaning 'fox-red', which might be a useful descriptive epithet, but not for the one plant that bears it (the aforementioned pale yellow fox-bane), and the more usual specific epithet for the fox is *vulpinus*. True fox sedge, *Carex vulpina*, is a threatened wetland species of grass with rust-coloured inflorescences that have made it attractive to gardeners. (The very similar false fox-sedge, however, is *Carex otrubae*, its name taken from Josef Otruba, a twentieth-century Czech postal worker who botanised in Moravia.) Similarly, the cactus *Rebutia vulpina* is named for its orange-red flowers. On the other hand, the North American *Vitis vulpina* appears to have been named by Linnaeus with Aesop's fable 'The Fox and the Grapes' in mind. Thwarted in

his repeated attempts to reach a bunch of grapes growing high on a vine, Aesop's fox proclaims the fruit sour in any case and so not worth his while – hence the term 'sour grapes'. When growing wild *V. vulpina* does indeed scramble up into trees, and its fruit is sour until it has been exposed to frost, which sweetens it; for this reason the plant is also known as the frost grape. Unhelpfully, and for no reason I can discover, it is also known as the chicken grape, while in America another species, *Vitis labrusca* (the epithet

Grape vine, *Vitis*

being the Latin word for a wild vine), is also known as the fox grape because of its 'foxy' musk, a technical term in winemaking that is more or less a synonym of 'earthy'.

The domesticated dog belongs to the same biological family as the wolf and the fox, the *Canidae*, and has also supplied botanists with a number of plant names. Who would have thought that alyssum, that favourite of the English rockery, owed its botanical name to dogs? Thought to be a cure for rabies, it was named from the Greek prefix *a-*, meaning against, and *lyssa* meaning rage or frenzy, and the plant was once known (along with several others) as madwort. The prefix *cyno-* is from the Greek word for dog, *kyon* (gen. *kynos*)– and is not to be confused with *cyano-*, from the Greek *kyanos*, meaning dark blue. It is most familiar to gardeners from *Cynoglossum*, the English name of which, hound's-tongue, is a direct translation of the Greek – *glossa* means tongue – and refers to the long shape and roughness to the touch of the plant's leaves. According to Gerard, these leaves have another canine property: they 'stinke very filthily, much like to the pisse of dogs; wherefore the Dutch men have called it Hounds pisse, and not Hounds tongue'. Geoffrey Grigson, on the other hand, suggests in *The Englishman's Flora* that it 'has a strong smell of mice', and records that the plant is known as rats-and-mice in Wiltshire. Regardless of this supposed smell, cultivated species such as *Cynoglossum amabile* have long been valued by gardeners for their bright blue flowers and long flowering season. Hound's-tongue seeds very freely, and in Gerard's time it grew 'almost every where by high-wayes and untoiled ground'. Parkinson evidently did not think it garden-worthy, but in recent years the dark maroon flowers of the original wild species, *C. officinale*, have become fashionable. At the other end of the dog, *Cynosurus cristatus* is a common British grass which takes its genus name from *kyon* + *oura*, the Greek for tail, and has the English name of crested dog's-tail. *Cynosurus echinatus* is a decorative grass that combines the dog and the hedgehog (*ekhinos* in Greek), further zoological confusion contributed by the RHS which calls it cock's comb grass, though it is also and more sensibly known as rough dog's-tail.

The word cynic comes from the same Greek root as 'dog', because Diogenes, one of the leading members of the Cynical school of philosophers, was criticised for living like a dog according to its needs – eating when hungry, sleeping when tired wherever one happened to be, and disregarding the niceties of normal social life. The Latin specific epithet *caninus* also tends to refer to the less admirable aspects of dogs' lives or their place in the lives of humans. In many cases it means no more than 'very common, as plentiful as dogs'; elsewhere it means inferior, as in dog Latin or doggerel. The dog rose, *Rosa canina*, was once a widespread countryside plant, and because it has single flowers with only a faint scent it was considered inferior to cultivated roses grown for their strong perfume and many–petalled blooms. That said, there can be few more gladdening sights than the variable pink or white flowers of this plant unexpectedly encountered in an English hedgerow. Sitting in the Café des Westens in Berlin in May 1912, a homesick Rupert Brooke compared the regimented tulips in the city's public gardens with dog roses back home: 'Unkempt about those hedges blows / The English unofficial rose'. Similarly, the dog violet, *Viola canina*, is a British native that has flowers that are as pretty as any violet in cultivation but are unscented. In case of canine confusion, it should be noted that *Viola labradorica* refers to a geographical location – it is a native of Canada – rather than a breed of dog, while the dog's tooth violet, *Erythronium dens-canis*, isn't a violet at all, or even in the same family, being a member of the *Liliaceae* or lily family. The reason it is called dog's tooth (which is simply a translation of the Latin specific epithet) is also hard to discern when you see the plant's delicate single flower-head rising from two spotted leaves. The clue is below ground: the bulb is white, long and pointed, reminding early gardeners of dog's teeth.

In his chapters about native British orchids, Gerard refers to a genus he calls *Cynosorchis*, for which he gives the common name Dogs stones or Dogs cullions, explaining: 'The roots be round like unto the stones of a Dog, or two olives, one hanging somewhat shorter than the other'. While we are dealing with the rear ends of dogs, we should mention *Plectranthus caninus*, which, like cynoglossum, smells strongly of dog's urine and as such is marketed (as *Coleus canina*) as a cat repellent. Unlike cats, humans have to rub the leaves to release the off-putting smell, and the plant has stubby lavender flower-spikes to compensate for its unsavoury reputation. Those who like cats and wish to provide them with a natural high, can grow *Nepeta*, a genus often given the blanket English name of catmint. The tall *Nepeta cataria* (from *catta*, a late Latin synonym for the more usual *feles*), which has been grown in English gardens since the thirteenth century, has pale pink flowers and is the true catnip, sending cats into rolling ecstasy, though other species have a similarly intoxicating effect.

Staying with cats, the king of the beasts (whose Latin binomial is *Panthera leo*) naturally lends his name to various plants either by way of colour or appearance, notably the teeth, tail and ears. Hawkbit gets its common name from a medieval notion, probably derived from Pliny, that falcons ate the plant to improve their sight. Its botanical name, however, is *Leontodon*, from the Greek for lion's tooth, probably because it is similar to the dandelion, the English name of which is a corruption of

the French *dent de lion*, itself
a translation of the old Latin
name *Dens leonis*. Both plants
have jagged-edged leaves and
composite flowers made up
of bright yellow florets, which
might have suggested their
names; they are, however,
botanically unrelated and the
genus name of the dandelion,
Taraxacum, is from the Persian
for 'bitter herb'. *Leonotis*
means lion's ear and is a
genus of South African plants,
of which the best known
is the anatomically confused
Leonotis leonurus, known as

Dandelion, *Taraxacum officinale*

lion's tail because its tall flower-stems have whorls of orange
flowers. *Leonurus* (from the Greek *leon* (gen. *leontos*) = lion +
oura = tail) is another genus, with the less than leonine
common name of motherwort. The specific epithet *leontoglossus* is derived from the Greek *leon* + *glossa* = tongue. Peering
into the spotted mouth of the yellow South African orchid
Orthochilus leontoglossus apparently gave some botanist a sense
of being a Christian in a Roman arena. *Leonensis* generally
means coloured like a lion, as in *Narcissus nobilis* var. *leonensis*,
which has a deep yellow corona, the colour extending to the
tepals, but *Penstemon leonensis* is so named because it was
discovered in 1934 in the Mexican state of Nuevo León. A
number of botanical dictionaries give *leonis* as an epithet
meaning coloured or toothed like a lion, even citing the white
orchid *Angraecum leonis*, but this plant was in fact named by

the English breeder Edith Watson after her father's employer Sir Herbert Leon of Bletchley Park (subsequently famous as the headquarters of the Second World War codebreakers).

Also in the cat family is the African civet, for a long time the principal source of another musk used in perfumery, including such famous brands as Chanel No. 5 – although the company now uses a synthetic version instead of the animal's pineal secretions. 'Give me an ounce of civet, good apothecary, to sweeten my imagination,' the mad King Lear cries in Shakespeare's play, but in *As You Like It* Touchstone refers to the perfume bluntly as 'the very uncleanly flux of a cat'. In its pure form the secretion (sometimes called zibeth) has a very strong and unpleasant smell, and it is for this reason that the notoriously stinking (but delicious) durian fruit has the botanical name *Durio zibethinus*.

A principal prey of the lion, the zebra, lends its name to distinctively striped plants such as the hothouse *Calathea zebrina*, though most gardeners will have more often come across this epithet used as a cultivar name, as in the green-and-gold-striped grass *Miscanthus sinensis* 'Zebrinus' or the old-fashioned mallow *Malva sylvestris* 'Zebrina', which has mauve flowers striped with dark purple veins. Also originating on the African plains is the humble donkey or ass, who in various fables is often depicted eating thistles rather than plants more usually considered edible. The *Onopordum* genus of thistles, which includes the stately grey-leaved cotton thistle *O. acanthium*, derives its name from the Greek *onos* = an ass + *porde* = a fart, this apparently being the result when donkeys gorge on the plant. Whether sainfoin has the same effect is unrecorded. It is sometimes thought that the plant's English name is a corruption of the French *saint foin* or 'holy hay' because of an old legend that it was present in the manger where Christ was born, but it is in fact derived from the Latin *sanum foenum*, which means merely healthful hay, while its botanical name, *Onobrychis*, takes the 'ass' prefix and adds it to *brycho*, meaning to eat greedily – as the ox and the ass no doubt did in their Bethlehem stable.

Both donkeys and zebras are members of the horse family and the prehistoric survival *Equisetum* takes its name from *equus*, the Latin for horse, and *saeta*, meaning bristle. Dismissed by Stearn as 'of more historical than gardening interest', some species of horsetail are grown by those who like horticultural curiosities and for a while became fashionable as a cut 'flower' to display in simple glass vases. The specific epithets *equinus* and *equestris* seem not much used, apart from the popular orchid *Phalaenopsis equestris*, though quite why (in 1843) the German botanist Johannes Schauer gave the name to this species of moth orchid is unclear. The Greek prefix for horse, *hippo-*, is more familiar, chiefly because used of a tree so much part of the British countryside that people imagine it a native species rather than an introduction from the time of Charles I. The horse chestnut, *Aesculus hippocastanum*, is not a chestnut at all, nor indeed (as Linnaeus, who named the genus, surmised) a kind of oak – though it is related to horses. The first reports of the tree, a native of the Balkans seen growing in cultivation in Istanbul in the mid-sixteenth century, recorded that conkers were fed to horses to relieve pulmonary complaints and worms. The second element of the specific epithet is derived from *kastanon*, the Greek word for the sweet chestnut, the fruit of which is similar to the conker in that it is glossy brown and contained within a prickly shell. There is no written record of the tree growing in the wild until the 1790s when a Cornish traveller called John Hawkins seems to have come across it in Greece. At some point before this, it had been decided that the horse chestnut was in fact a native of India, and this erroneous belief is commemorated in the French and Italian names for the tree, *marronier d'Inde* and *castagno d'India*. There is in fact an Indian horse chestnut, *Aesculus indica*, which is native to the Himalayas, but grown in British gardens for its long pink 'candles' of flowers.

The possibly equine origins of the hippeastrum's name are more complex. It was the botanist, clergyman and MP William Herbert (1778–1847) who separated this South American bulb from the similar but South African *Amaryllis* – not that everyone took much notice, and some two hundred years later hippeastrum bulbs are still sold in gift-boxes as 'amaryllis'. Stearn, clearly sceptical, quoted a late eighteenth-century botanical magazine which recorded that the spathe or sheath containing the flower buds 'is composed of two leaves [sic], which standing up at a certain period of the plant's flowering like ears give the whole flower a fancied resemblance to a horse's head'. One's fancy would have to be very well developed to think this, but other suggested derivations are not a great deal more helpful. One is that the root of the name is not *hippos* but *hippeus*, a rider, and that 'the equitant leaves suggest being astride a horse'. 'Equitant' is a botanical term meaning that leaves overlap or stand one within the other, and although the word is clearly derived from *equus* and related to equitation, it seems a bit of a stretch to see this arrangement of leaves as horse-straddling. The rest of the genus name is derived from the Greek *astron*, so the plant's name in full might be translated as horse (or rider) star, presumably for the flowers, which are in fact trumpet-shaped.

A genus name once used widely but long since discarded is *Hippoglossum*, or horse-tongue, for the *Ruscus* genus of tough, evergreen, red-berried shrubs. It is now more commonly called butcher's broom, supposedly because branches of it were used to sweep meat counters and chopping blocks. The viciously pointed 'leaves' of the true butcher's broom (in fact shoots that take a flat leaf shape) are acknowledged in its botanical name, *R. aculeatus*, meaning prickly. Other animal tongues include those of cattle in the genus *Buglossoides*, from the Greek *bous*, meaning ox. These plants do have long leaves, but the ones most often grown in gardens for their bright blue flowers have been moved to different genera: *Lithospermum* (from the Greek *lithos* = stone + *sperma* = seed) and *Lithodora* (*doron* = gift, referring to the fact that in the wild they emerge from rocky habitats). All these plants belong to the borage family, notable for the roughness and hairiness of their leaves, and the tongues of cattle came into the same category as those of dogs and horses. A number of plants in this family are still called bugloss, common bugloss being a synonym for alkanet or *Anchusa officinalis*, Italian bugloss for *Anchusa azurea*, Siberian bugloss for *Brunnera macrophylla*, and the zoologically confusing viper's bugloss for *Echium vulgare*. Meanwhile *Buphthalmum* means ox-eye, and *B. salicifolia* is the yellow ox-eye daisy, odd names perhaps explained by the central boss of the flowers which could be thought to resemble the slightly bulbous eyes characteristic of cattle. It is the central rib of the leaves of some kinds of bupleurum (*bous* + *pleura* = rib) that, with some exaggeration, justify that genus's botanical name. The English name for these plants, hare's ears, sensibly emphasises the more noticeable length of the leaves. Other English plant names drawn from cattle include cowslip and oxlip, which sound pretty enough but in fact refer to cow's dung (*cuslippe* or *cusloppe* in Old English), since it was thought these plants sprang up from cowpats. Both plants are primulas, and their botanical names have nothing to do with animal husbandry. *Vaccaria hispanica*, however,

appears to derive its name from *vacca*, the Latin for cow, probably because it was used to fodder cattle. The reason this pretty
pink-flowered member of the pink family was fed to cows is
that it is supposed to promote lactation, and it has the common
names of cowherb and cow cockle.

Goat's rue, *Galega officinalis*, was another fodder crop
intended to increase milk yields, and the plant's genus name
is derived from the Greek *gala* (milk) + *aigos* = goat. Similarly
derived is the genus name of that very unwelcome garden vis-
itor ground elder, *Aegopodium podagraria*, which in Northern
Ireland has the popular name of goat's foot, a direct translation
of the Greek. Another Greek word for a goat is *tragos*, as in
Tragopogon pratensis, usually known as Jack-go-to-bed-at-noon
(see 'Miscellanea') but also as goat's beard because of its long
silken hairs, *pogon* being the Greek for beard. Honeysuckle,
meanwhile, was named by
the sixteenth-century German
botanist Adam Lonitzer
Lonicera caprifolia from the
Latin for goat (*caper* = billygoat / *capra* = nannygoat),
because these animals enjoyed
eating the plant's leaves. It is
possible that the species name
came from European common
names already in existence,
many of which reference goats:
Geißblat in Lonitzer's own language, *kaprifol* in Linnaeus's
Swedish, *caprifoglio* in Italian,
chèvrefeuille in French. Given
that goats will eat almost

Goat's beard, *Tragopogon pratensis*

anything, it seems odd to single out this plant, but the *Caprifoliacea* family contains forty-two genera, including *Abelia*, *Leycesteria* and *Weigela*, all of which might provide a passing goat with something to nibble at. There is another species name, *tragophyllus*, which means precisely the same thing in Greek and appears now to occur only in another species of lonicera, the yellow-flowered (and unscented) *L. tragophylla* or Chinese honeysuckle, introduced to English gardens in 1900. One final plant related to goats is *Salix caprea*. It is sometimes known as pussy-willow and the gosling tree because of its fluffy catkins, but it is also called goat willow and owes both this name and its botanical one to the fact that, like honeysuckle, it was something herders encouraged their goats to feed upon.

The specific epithet *ovinus* from *ovis*, the Latin word for sheep, has probably been applied to plants for similar reasons. The extremely prickly blue devil, *Eryngium ovinum*, looks more than a match for any sheep, but apparently the vast flocks that graze in New South Wales, where the plant is a native, will eat it when it is young. Perhaps more generally appetising are the grass called sheep's fescue, *Festuca ovina*, the North American sheep's fleabane, *Erigeron ovinus*, and the somewhat obscure species of milk-vetch *Astragalus ovinus*.

Serpents are not much referred to in botanical names, but the aforementioned viper's bugloss, *Echium vulgare*, got its genus name from Dioscorides, who thought the seeds resembled the head of a snake and so borrowed from the ancient Greek word for viper, *ekhidna*. Despite the fact that a spiny ant-eating mammal, the echidna, has almost exactly this name, it should not be confused with the Greek word for a hedgehog or sea-urchin, which is *ekhinos*. Naturally enough, the prefix *echino-* is used for many kinds of cactus, but also for other kinds of plants with prickly flower-heads, including the invasive cockspur grass *Echinochloa crus-galli*. The spiny central boss that gives coneflowers their English name is referred to in their botanical name of *Echinacea*, while the combination of *echino*- with *-ops*, meaning 'like' results in *Echinops*, the electric-blue globe-thistle. Several specific epithets liken plants to hedgehogs. *Pelargonium echinatum*, for example, is named for its bulbous, cactus-like stems. The dune daisy, native to South Africa, is *Felicia echinata* because of the prickly appearance of its flower stems, which are densely covered with small leaves. *Agave echinoides* is named for the extremely sharp, dark brown tips to its leaves (a colouring reflected in its recategorisation as *A. striata*); and the shrub tan oak, *Lithocarpus densiflorus* var. *echinoides*, and *Euonymus echinatus* are both named for their toothed leaves. The needle-like leaves of some conifers also suggest hedgehogs: despite being known as the short-leaf pine, *Pinus echinata* has dense bundles of rather long needles, while the white spruce, *Picea glauca* 'Echiniformis' [sic], grows into a low hedgehog-like mound of needles. *Ceratophyllum echinatum*, unhelpfully known as spineless hornwort, is a familiar whorled-leaved pond-weed often used to oxygenate water. The scientific name for the hedgehog, *Erinaceus*, is used as an additional specific epithet relating to that creature. *Dianthus erinaceus* is a rockery pink that forms dense little domes of spiny foliage studded with pink flowers, while the Mojave prickly pear is *Opuntia erinacea*, but is also known – in yet another zoological muddle – as the Grizzly Bear cactus.

Bears roamed the forests of Europe when many plants were first named and are invoked in such names as bearwort for the aromatic umbellifer *Meum athamanticum* (otherwise spignel), bears' ears for auriculas (because of the shape of their leaves), or bears-breech for acanthus. The last name is sometimes rendered bears' breeches, suggesting ursine trousers, but this is merely a polite form of the correct original, 'breech' meaning backside or hind-quarters, as in breech-birth. No feature of the plant explains this name, but it was also known as Brank-ursine, which is from the medieval Latin *branca ursina*, meaning bear's claws, and refers to the curved shape of the flower as it emerges from the calyces the way an animal's claw does from its sheath. (Bear's claws, unlike those of cats, are not in fact retractable.) None of these plants carry over the association into a botanical name, but bears have their own specific epithet, *ursinus*, which supposedly means several things: shaggy in appearance, northern (from the Great Bear constellation), or attracting bears. It is this third that is usually intended, and the most recognisable plant with the name is wild garlic, or ransoms, *Allium ursinum*, the bulbs of which are often dug up by bears searching woodland for food. The California blackberry, *Rubus ursinus*, is also popular with foraging bears. The genus of vibrant orange and yellow African daisies *Ursinia*, however, has nothing to do with bears, but is named for the German botanical author Johannes Heinrich Ursinus (1608–67).

The delphinium owes its name to a rather more gentle animal than the bear, *delphis* being the Greek for dolphin, *delphinus* the Latin. The origin of the name of the creature itself is intriguing. According to a twelfth-century bestiary translated by T. H. White, the animals were called this 'because they follow the human voice or else because they will assemble together in schools for a symphony concert'. The first explanation refers to the human voice of the Oracle at Delphi, which attracted those who wanted to listen to her prophecies, and to the legend that Apollo first came there in the form of a dolphin. The second refers to an ancient belief that dolphins danced on the waves in response to music – they were sometimes referred to as *philomousoi* or music-lovers. The seventh-century B C musician Arion supposedly escaped from pirates by playing his lyre aboard their ship, summoning dolphins who bore him on their backs to the safety of the shore. As for the resemblance of delphinium flowers to these sea creatures, Gerard notes of the larkspur, that more modest, annual relative of the tall border perennial, 'the floures, and especially before they be per-fected, have a certain shew and likenesse of those Dolphines, which old pictures and armes of certain antient families have expressed with a crooked and bending figure of shape'. In other words, we should be thinking not of the sleek creatures now familiar to us on television or at sea aquariums, but of the ones sometimes depicted in the corners of old maps, or the cast-iron ones curving round the lamp standards along the Thames Embankment in London. The flower-spur that sug-gested a dolphin's tail has been largely bred out in the modern, double-flowered delphinium, although may still be discerned in the flower bud.

Larkspur, *Delphinium ajacis*

Corydalis, which has flowers somewhat similar to larkspur, takes its name from the Greek *korydallis*, meaning 'crested lark', and the larkspur itself was also known in Gerard's time as Larkes heel, Larkes claw and Larkes toe, and in France as *pied d'alouette*. The specific epi-thet *ornithopodus* is the Greek for bird's foot, and usually restricted to grasses such as the bird's-foot sedge, *Carex ornithopoda*, the flower-spikes of which separate out to resemble three birdlike 'toes'. The feet of a specific bird give their name to the dreary-looking but edible and intriguingly named plants Fat Hen (*Chenopodium album*) and Good King Henry (*C. bonus-henricus*). The genus takes its name from the Greek for goose foot, a reference to the shape of the plants' leaves. The Latin word for goose is *anser*, giving rise to the specific epithet *anserinus*. This name has nothing to do with leaf shape, but supposedly refers to plants that were either fed to geese or grew where geese were kept. The botanical name of the yellow-flowered silverweed, so called because of the grey backs of its leaves, is *Potentilla anserina*, and in Sweden it is called goosewort, presumably for one of these reasons. However, *Eriogonum strictum* var. *anserinum*, with its delicate umbels of yellow flowers on top of long stems rising from a mat of silvery leaves, is known as Goose Lake wild buckwheat and takes its botanical name from a place rather than from the bird itself.

The specific epithet *avicularis* is at the other end of the ornithological scale from the goose, derived from *avicula*, a small bird. This does not really explain the naming of *Polygonum aviculare*, but gardeners will hardly care since this plant is the invasive and more or less ineradicable common knotgrass. *Avicula* is a diminutive form of the standard Latin word for bird, *avis*, the genitive plural of which is *avium*, as in the wild cherry or gean, *Prunus avium*. This is the cherry that A. E. Housman declared 'loveliest of trees' in *A Shropshire Lad* because of its white blossom, but the fruit is thought fit only for birds, hence its botanical name. The common name of bird-cherry, however, refers to *Prunus padus*, the species name being what Theophrastus called this tree. Although birds seem not particularly attracted to the temptingly bright red berries of rowan or mountain ash, the tree is named *Sorbus aucuparia*, from the Latin *avis* + *capere*, to catch. Ancient bird-catchers apparently used rowanberries as bait, incorporating them in birdlime, so perhaps birds' tastes have changed down the centuries, or they have now become more fussy with access to an increased variety of available fruits. The star of Bethlehem genus also owes its botanical name to bird-catchers. *Ornithogalum* is the Latinised version of a Greek word meaning 'bird milk' and was coined by Dioscorides because the sap or some other part of the plant was used as birdlime.

The name of the bright yellow celandine, familiar from woodlands and verges but also grown in gardens, derives from a specific bird. Garden plants of deeper shades, or with cream, white or double flowers, are generally cultivars of the lesser celandine, *Ranunculus ficaria*, but the greater celandine, *Chelidonium majus*, takes its genus name from the Greek word *chelidon*, meaning the swallow, since the plant was traditionally believed to come into flower at the same time that these birds arrived from Africa to spend the summer in Europe. The genus name of the hawkweed, several species of which have become garden plants, is *Hieracium*, derived from the Greek *hierax*, a hawk. Pliny related a story of hawks using the juice of a plant of this name to sharpen their sight, a notion related to the medieval one that gave hawkbit its common name, and Gerard suggested that humans may similarly benefit 'if the juice of them be dropped into the eyes'.

The birds from which geraniums, pelargoniums and erodiums take their names have already been mentioned in the first chapter, but the long-stalked cranesbill, *Geranium columbinum*, adds another bird to the mix. *Columba* is the Latin word for a dove or pigeon, and the deeply lobed leaves of this plant bear some resemblance to the feet of these birds. (Confusingly, dove's foot cranesbill is the English name given to another plant altogether, the invasive but thankfully annual weed *Geranium molle*.) The columbine too takes its English name from doves, not (as I once thought) from the *Commedia dell' arte* character. Indeed, 'columbine' was used for the plant as early as the thirteenth century, three hundred years before Harlequin first sighed. Gerard explains that 'the slender sprigs [on the flowerstalks] bring forth every one one floure with five little hollow hornes, as it were hanging forth, with small leaves hanging upright, of the shape of little birds'. If you look closely at a single variety you can see what he means: the upcurving spurs look like the necks and heads of a cluster of birds, and the petals (as Tournefort puts it) like 'expanded Wings'. Earlier botanists saw instead the claws of a more fearsome bird, naming the genus *Aquilegia* from *aquila*, the Latin word for eagle. This

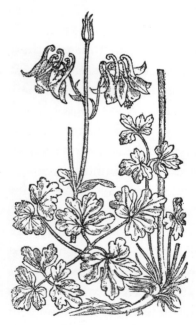

ornithological muddle sounds like an Aesop fable, which it isn't, but 'The Eagle and the Dove' is the title of a story in the Buddhist sutras, a poem by Wordsworth and a joint biography of two saints by the gardening writer V. Sackville-West. The eagle also lends its name to the common holly, *Ilex aquifolium*, the specific epithet referring to the hooked, beak-like spines of the leaves. Similarly prickly plants that share this specific epithet include *Mahonia*, *Berberis*, *Grevillea* and

Columbine, *Aquilegia*

Osmanthus. Linnaeus named common bracken *Pteridium aquilinum* not because of its large, wing-like leaves but because he saw the image of an eagle when looking at a transverse section of the plant's root – another example of a botanical name quite literally buried.

Bracken is of course a kind of fern, many of which take the species name *scolopendrius*, from the Greek word for millipede, because the pattern of the spore patches on the backs of the leaves are reminiscent of this many-legged creature: the monarch fern is *Phymatosorus scolopendria*, while (in yet another zoological disagreement between Latin and common names), the hart's tongue fern is *Asplenium scolopendrium*. At least there is linguistic agreement about the ostrich fern, which is *Matteuccia struthiopteris*, from the Greek *strouthos* = ostrich + *pteron* = feather. *Strouthos* is in fact the Greek name for a sparrow, but the ancient Greeks were at a loss to explain a bird as large as an ostrich and so called it a sparrow-camel, in just the same way that the giraffe was called a camel-pard (*kamelos* + *pardos*), a name that as 'cameleopard' persisted in English into the Middle Ages. Someone must have very sensibly decided that *struthiocamelopteris* would have been a bit of a mouthful, and so the ostrich part of the fern's name was shortened, presumably causing some confusion to ornithologists trained in the classics. The large feathery leaves of the plant have a faint resemblance to ostrich feathers, and are arranged in a manner that gives rise to the alternative common name of shuttlecock fern.

Like the hart's tongue fern, the English name of the snake's-head fritillary is at zoological odds with its Latin one, *Fritillaria meleagris*, which also owes its specific epithet to a bird. According to Geoffrey Grigson, the English name is of only nineteenth-century origin, just in time for the poet and

Snake's-head fritillary,
Fritillaria meleagris

keen-eyed amateur naturalist Gerard Manley Hopkins to note: 'Buds pointed and like snakes' heads, but the reason of name from mottling and scaly look.' *Fritillaria* comes from *fritillus*, the Latin word for dice-box, an object most usually used in conjunction with a chequered games-board – and the French call the plant *damier*, which means a draughts-board. *Meleagris* derives from the Greek word for guineafowl, the specked feathers of which are very like the distinctive, dusky purple chequerboard markings of the flower. 'The first time I saw guineafowl,' recalled Grigson's wife Jane, the cookery writer and food historian, 'they were humped along the roof ridge of a French farmhouse, like a row of black and white chequered tea cosies.' In Gerard's time they were known as 'the Ginny-hen floure', a name which certainly survived into the 1950s in some parts of Britain. The markings of the snake's-head fritillary have always struck people with wonder, Gerard piously declaring that they show how 'Nature, or rathere the Creator of all things, hath kept a very wonderfull order, surpassing (as in all other things) the curiousest painting that Art can set downe.'

Anemones filled early gardeners with almost as much wonder as fritillaries, chiefly because they seem to come in such a wide variety of colours. It is perhaps unsurprising, therefore, that one species should be named after that showiest of birds, the peacock. *Anemone pavonina* derives its name from *pavo* (gen. *pavonis*), the Latin name for the bird that not only symbolised the goddess Juno but often ended up on the Roman dinner table, roasted but with its iridescent tail feathers reattached to form a glittering centrepiece. The astonishing light-reflecting metallic blue leaves of *Begonia pavonina* evidently suggested these feathers to Henry Ridley (1855–1956), an English botanist working in the Malay Peninsula who first named the plant; but it was probably just the fanned leaves that led Linnaeus to name the Asian coral-bean tree *Adenanthera pavonina*. The annual peacock poppy, *Papaver pavoninum*, has flowers of a bright blood red, a colour wholly absent from the bird's feathers, but the name refers to the dark blotch at the centre, which was thought to resemble the 'eye' of a peacock's tail feather. (The *Pavonia* genus of mallows, incidentally, has nothing to do with peacocks but is named after the Spanish botanist José Antonio Pavón, who collected plants in South America for the Real Jardín Botánico de Madrid between 1778 and 1788.)

The similarly colourful plumage of parrots led to the specific epithet *psittacinus*, derived from the Latin name for the bird, *psittacus*. The red and green South American *Alstroemeria psittacina* is known as the parrot lily, and these two colours, reminiscent of the red and green macaw, also distinguish *Hippeastrum psittacinum*, whereas the flowers of the tropical shrub *Cestrum psittacinum* are the vivid saffron of the blue and yellow macaw's breast feathers. The parrot or cockatoo balsam, *Impatiens psittacina*, was discovered in Burma in 1899 by the civil servant and plant collector A. H. Hildebrand (1852–1915), who described it as 'a pretty, compact plant [...] covered with flowers which resemble a Cockatoo suspended by a string from the shoulders'. The species name of *Heliconia psittacorum* is the genitive plural of *psittacus*, meaning 'of parrots', and refers to the plant's bright red upright flowers, which also give it the common name of parrot's beak. It should be noted in passing that despite its red flowers and colourful autumn foliage, the Persian ironwood, *Parrotia persica*, has nothing to do with the *Psittacidae*, but is named for Friedrich Parrot (1791–1841), a Baltic German naturalist and mountaineer.

A cock's spur is not technically a claw, hut a horny growth on the back of a fowl's leg, and the specific epithet *crus-galli* (from the Latin for cock's leg) has already been mentioned for cockspur grass. More familiar, perhaps, is the hawthorn *Crataegus crus-galli*, named because the tree's large thorns resemble cocks' spurs. The Latin word for a cockscomb is *crista* (from which we get the word crest), but *Erythrina crista-galli* is known as the cockspur coral tree, a name evidently coined by someone who didn't know one end of a bird from the other. Presumably the Latin name refers to the flowers, which resemble a cockscomb in that they are red and waxy, but in shape clearly mark the plants as a member of the bean family. At least the brightly coloured crests of another cockscomb, *Celosia cristata*, live up to the English name as well as having a botanical name that makes a good deal more sense than the erythrina's. That said, on its own *cristata* means no more than crested, with no ornithological connotations – as in *Iris cristata* or the curiously corrugated cactus-like *Euphorbia lactaea* f. *cristata*.

This euphorbia is sometimes known as crested elk-horn, and plants with branched leaves resembling such antlers have their own specific epithet, *alcicornis*, from the Latin *alces* = elk + *cornu* = horn. Inevitably, a euphorbia not called elk-horn bears the botanical name *Euphorbia alcicornis*, while the staghorn fern is *Platycerium alcicorne*. Little horns take the Latin diminutive, and so bird's-foot trefoil is *Lotus corniculatus* because it has horned seed capsules. The creeping wood sorrel *Oxalis corniculata* is named for the same reason, and although attractive in its purple-leafed variety, is simply impossible to get rid of once you have it in the garden, where its roots insinuate their irretractable way among those of any other plant in the vicinity. The Latin word for a ram, *aries*, is familiar from the zodiac, and *arietinus* means with the head or horns of a male sheep. The orchid *Cypripedium arietinum* has the rather absurd English name of the ram's head lady's slipper, which sounds like something out of a pantomime, while *Paeonia mascula* subsp. *arietina* presumably got its name from the curved shape of the flower's carpels. And if you have ever seen the white curved shoot emerging from a sprouting chickpea, you will understand why the plant's botanical name is *Cicer arietinum*.

The Latin words for mouse (*mus*) and fly (*musca*) are close enough to cause some confusion when used in botanical names. The English name for the bright-red spotted toadstool familiar from fairytales, for example, is more or less a translation of the name given to it by Linnaeus: *Agaricus muscaricus*, the fly agaric. The genus name was changed not long afterwards to *Amanita* (inexplicably from the Latin *amans* meaning lover), but the fly-attracting specific epithet was retained and feminised as *muscaria*. You might think that the name of the insectivorous Venus flytrap, *Dionaea muscipula*, would be similarly derived, but the specific epithet refers to the plant's mechanism rather than its prey, *muscipula* being the Latin for mousetrap. Ah yes, you may say, I can see the difference between these two epithets, but still wonder what the grape hyacinth has to do with flies or mice. The answer is nothing: *Muscari* is not derived from Latin but from the Greek word *muschos* meaning musk, an allusion to the powerful scent of this little plant, so cheering on a cold spring day. As its doubled name suggests, the purple and buff *Muscari moschatum* is renowned for having the best scent of all grape hyacinths.

The musk flower *Mimulus moschatus*, meanwhile, takes its genus name from the Greek word for ape, *mimo* – or possibly the Latin word for actor, *mimus*. The two words are connected, and not only in classical languages: to 'ape' something or someone is to imitate, as an actor does. Other species of mimulus are called monkey flowers, and the imaginative have claimed to see a monkey's face in the seed or corolla. Those who prefer the theatrical derivation suggest that the flower resembles the kind of mask worn by actors in ancient Greek theatre. Peer into the flower of the orchid *Dracula simia*, however, and you will see a monkey's face very distinctly, *simius* being the Latin for monkey. The genus of this plant is named for Bram Stoker's Count Dracula, whose bloodsucking exploits kept the Hammer film studios in business for many years, and refers both to the elongated 'wings' of the flowers, reminiscent of a bat, and to their often dark colour, ranging from venous red to near black.

The frequently curious flower structures of orchids either look like, or are intended to imitate, other things, and it is the dangling, manikin-like three-lobed lip that gives the European monkey orchid, *Orchis simia*, its name. As Jocelyn Brooke writes in his authoritative monograph *The Wild Orchids of Britain*: 'The supposed resemblance to a monkey, if not immediately striking, is at least easier to detect than the soldierly qualities of *O. militaris*, and to the eye of faith there is, it is true, a somewhat simian and prehensile quality about the slender, curling segments of the lip.' It is the long, pale green lip of the lizard orchid, *Himantoglossum hircinum*, that gives it its English name, as well as the genus name, which is the Greek for strap-tongued. The species name, however, refers to another animal altogether, *hircus* being another Latin name for the billygoat. This epithet has nothing to do with the plant's appearance, but refers to what Gerard calls its 'ranke or stinking smell or savour like the smell of a Goat' – though for one of Brooke's friends, the smell was instead a horrifying reminder of the miasma hanging over the battlefields of the First World War. Of the frog orchid, Brooke writes: 'Once again, one must enlist the eye of faith to detect any batrachian quality' in the plant, although this is somewhat belied by the beautiful watercolour of the plant Stephen Bone provided for Brooke's monograph, in which the individual flowers have a certain frog-like appearance. The botanical name, however, honours the green colour of the flowers rather than their shape: *Coeloglossum viride*. (The genus name, incidentally, is Greek for hollow-tongued, and should not be confused with the Latin prefix *coeli-*, referring to the sky or heavens and usually denoting a blue colour.)

Among other European orchids is the bee orchid, *Ophrys apifera* (from the Latin *apis* = bee + *fero* = to bear), so named because the flower's hairy lip looks astonishingly like a bee, as does that of *O. bombyliflora*, which takes its species name from the bumblebee genus, *Bombus*, which is the Latin word for 'buzzing'. Altogether more complicated is the name of the two European spider orchids. The early spider orchid is *Ophrys sphegodes*, and the late spider orchid *Ophrys fuciflora*, but their specific epithets do not refer to spiders at all. The plants were, however, originally classified as *Arachnites*, a now defunct genus that took its name from *arakhne*, the Greek for spider. (It now survives as the specific epithet for a very spidery species of the *Aeranthes* genus of tropical orchids.) An entomological rethink led to the early and late spider orchids taking their species names respectively from the Greek word for wasp, *sphex*, and the Latin name for a drone bee, *fucus* – although the seemingly inflated lip of the former in particular looks very like the body of a large spider. The specific epithets *arachnoideus* and *araneosus* (from *araneum*, the Latin for spider's web) usually refer to plants covered with fine hairs, such as the cobweb houseleek, *Sempervivum arachnoideum* or the Brazilian hardy gloxinia, *Sinningia araneosa*.

The name of the moth orchid, *Phalaenopsis*, is derived from *Phalaena*, a now discarded genus created by Linnaeus to include all members of the *Lepidoptera* order apart from butterflies and hawk moths. The botanical names of the greater and lesser butterfly orchids, *Platanthera chlorantha* and *P. bifolia*, however, have nothing to do with lepidoptery but refer to flowers and leaves. The pink butterfly orchid *Anacamptis papilionacea*, however, takes its species name from the Latin word for butterfly, *papilio*. There is a species of *Moraea*, a lovely genus sometimes known as the Cape tulip but in fact looking like an iris and indeed belonging to that family, called *Moraea papilionacea*, but it has no particular claim above others in the populous genus to looking like a butterfly. With its two upper petals suggesting to the suggestible a pink butterfly, *Pelargonium papilionaceum* perhaps earns its name, and the massed leaves of *Oxalis triangularis* subsp. *papilionacea*, especially the deep purple ones of the popular 'Atropurpurea' cultivar, can give the appearance of a cluster of resting butterflies. Similarly the American common blue violet was formerly *Viola papilionacea* because the two upper and two lower petals of its flowers really do resemble butterfly's wings, while the downward-pointing fifth petal stands in visually for the thorax. It has now been renamed *V. sororia*, an epithet from the Latin word for sister and usually meaning very closely related. This also seems appropriate since in the early decades of the twentieth century the flowers became tokens of lesbian love on the grounds that in one of her poems Sappho depicts herself wearing a garland of violets. The other specific epithet from butterfly is simply the aforementioned Latin name for this insect, *papilio*. It might be conceded that (from a considerable distance) the drooping flowers of *Gladiolus papilio*, particularly in their softer shades, could be mistaken for butterflies hanging beneath the plant's stem, but the flowers of *Hippeastrum papilio* look no more like butterflies than those of any other hippeastrum – in other words, not in the least.

Butterflies are welcome visitors to the flower garden, but some plants owe their names to their supposed efficacy in warding off annoying insects. *Cimicifuga* is derived from the Latin words *cimex* (gen. *cimicis*), a bed-bug, and *fugare*, to drive away, and the English name of the plant is merely a translation: bugbane. The reason *Cimicifuga simplex* is now grown in gardens is not to keep bed-bugs at bay but because its tall flower-spikes and handsome foliage, particularly in the dark-leaved 'Atropurpurea' group of cultivars, add a dramatic touch to any planting scheme. Furthermore, the plant's vaunted insect-deterring properties were dismissed when its genus name became redundant. Bugbane is now classified as an *Actaea*, a genus that includes the highly poisonous baneberry, *A. rubra*, a plant so toxic it would deter creatures far larger and more destructive than bedbugs, two of its enticing red berries apparently being sufficient to kill a child. *Actaea*, incidentally, has a muddled etymology, taken by Linnaeus from Pliny, who saw a similarity between the leaves of this plant and those of the elder, for which the Greek name is *aktea* – though the two plants are not of the same family or even the same order.

Certain herbs were used to banish fleas, an even more

frequent household hazard than bedbugs. The Latin for flea is *pulex*, giving us the specific epithets of *pulegius* and *pulegioides*. Pennyroyal has a very strong smell which is supposed to repel even mosquitoes, and in the ancient world it was used against fleas and so has the botanical name *Mentha pulegium*. Broad-leaved thyme, *Thymus pulegioides*, had a similar reputation, and several other plants have the common name of fleabane. While the false fleabane, *Pulicaria*, retains the Latin root, *Erigeron*, the genus of daisies we most gener-

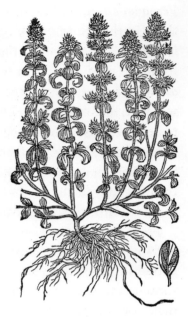

Broad-leaved thyme,
Thymus pulegioides

ally call fleabane, does not. Gerard called *Pulicaria* Flea-wort, 'not because it killeth fleas, but because the seeds are like fleas', whereas he classified what we now call erigerons in the *Conyza* genus of plants sometimes called horseweed and dismissed by Stearn as 'of no garden interest'. Gerard claimed that these plants 'burned, where flies, Gnats, fleas, or any venomous things are, doth drive them away'. Other plants take their name from their supposed efficacy in expelling worms, for which the specific epithet is *anthelminticus*, from the Greek *ant-* (against) and *helmins* (an intestinal worm). *Chenopodium anthelminticum*, known as American wormweed or, rather more enticingly, Mexican tea, is a plant of more interest to homeopaths than gardeners, but among the tall ironweeds whose bright purple flower-heads are an asset to any border is

Vernonia anthelmintica, a plant native to India and still in use there in the treatment of intestinal threadworms. Similarly, feverfew, *Tanacetum parthenium*, takes its botanical name from the medieval Latin word *tanazita*, itself derived from *athanasia*, the Greek word for immortality, because the plant was inserted in winding sheets in a presumably vain attempt to deter worms from getting to work on corpses. As its English name suggests, however, the plant was also used for more immediate bodily concerns, notably those at the other end of the life cycle, and its original genus name was *Matricaria* from *matrix*, a late Latin word for the womb.

From bears to intestinal worms, by way of numerous birds and insects, the natural world has inspired botanists to some extraordinary flights of fancy. They have had a similar effect on those botanical dictionaries that list mythical or misleading inclusions, some of which have already been mentioned. While the New Zealand yellow rock daisy, *Brachyglottis lagopus*, and the Mediterranean plantain, *Plantago lagopus*, take their species name from a hare's foot (from the Greek *lagos* = hare + *pous* = foot), I was disappointed to find no plant with the species name *lagocephalus*, apparently meaning with a head like a hare. It is hard to imagine what such a plant might look like – the plumed French lavender *Lavandula stoechas* comes to mind – but the *Artemisia lagocephalus* offered in one dictionary appears to be an accidental internet hybrid. An extract of *Artemisia campestris* has been trialled as a protection against poisoning by the liver of the oceanic puffer-fish, *Lagocephalus lagocephalus*, and an online search appears to have resulted in the taxonomic confusion between plant and fish. The epithet *camelinus* is no longer used to mean camel-coloured – if it ever was – and the genus name *Camelina* rather boringly derives from the Greek word for ground-flax. The more enticing *picus*, taken from the Latin for magpie or jay (*pica*) or woodpecker (*picus*) and meaning ornate, seems to have vanished, while *picaceus* is restricted to the beautifully mottled black and white magpie fungus, *Coprinopsis picacea*. Meanwhile, *porcinus*, from *porcus* the Latin for pig, appears to have abandoned the realm of botany for that of microbiology. The loss of such names seems to me to impoverish the language of gardening, but the one whose disappearance I regret most is the enchanting *vaccinus*, meaning 'the colour of a dun cow'.

Word List

alcicornis = antler-shaped, like the horn of an elk

anguilliformis = eel-like

anguinus = serpentine

anserinus = related to geese

anthelminticus = worm expelling

apianus = related to bees

apiferus = bee-bearing

aqui- (prefix) = eagle-like

arachnoideus, araneosus = cobwebbed

arietinus = like a ram's head or horns

aucuparius = bird catching

avicularis = relating to small birds

avium = of birds

bufonius = relating to toads or damp places

caninus = relating to dogs, very common

capreus = of goats

caprifolius = with leaves favoured by goats

cataria = related to cats

colubrinus = snake-shaped

columbinus = dove-like

corniculatus = with small horns

crista-galli = cockscomb

crus-galli = cock's spur

cyn- (prefix) = related to dogs

Cynoglossum = dog's tongue

cynorrhizus = dog's tail

dens-canis = dog's tooth

echino- (prefix) = hedgehog or sea urchin

elephantipes = shaped like an elephant's foot

erinaceus = hedgehog-like

gruinus = stork-like

hippo- (prefix) = related to horses

lagopus = hare's foot

leo-chromus = lion-coloured

leonensis = lion-coloured

leontoglossus = lion-throated

leonurus = like a lion's tail

leopardinus = strongly spotted

lupulus = small wolf

lycoctonus = wolf-killing

meleagris = guinea-fowl

murinus = related to mice

muscarius = relating to flies

ophiocarpus = snakelike fruits

ornithopodus = bird's-foot
ovinus = related to sheep
papilio = butterfly
papilionaceus = relating to butterflies
pardalinus = related to leopards
pavonina = related to peacocks
perdicarius = related to partridges
psittaceus = related to parrots
psittacorum = of parrots
pulegius, pulegioides = flea banishing

scolopendrius = relating to millipedes
scorpioides = curved like a scorpion's tail
sphegodes = wasp-like
struthiopteris = like ostrich feathers
tragophyllus = with leaves favoured by goats
ursinus = related to bears
vulpinus = related to foxes
zebrinus = zebra-striped
zibethinus = related to civet cats (smell)

EPONYMS

To most gardeners the names Leonhart Fuchs, Jean Rodolphe Lavater and Anders Dahl will be unfamiliar. Yet we all know and often refer to the plants named after these men, who were, respectively, a sixteenth-century German doctor and author, a seventeenth-century Swiss physician, and an eighteenth-century Swedish botanist. While some botanists and collectors gave their own names to plants they had discovered, many more – like these three men – had plants named in their honour by other people. Fuchs had been dead for well over a century when the French botanist and monk Charles Plumier discovered the fuchsia in Hispaniola and named it after him. Similarly, Dahl played no part in the discovery or introduction of the dahlia: it was named for him in 1791, two years after his death, by the Spanish botanist, clergyman and Director of the Madrid Royal Gardens Antonio José Cavanilles, who had been sent specimens of the plant by his opposite number in Mexico City. The origin of the lavatera's name is even more complicated, with some experts asserting that Joseph Pitton de Tournefort was honouring not Jean Rodolphe Lavater but a pair of naturalist brothers, Johann Heinrich and Johann Jacob Lavater. It seems Lavater was a not uncommon Swiss name in the seventeenth century, borne by several other distinguished men; but written evidence that Tournefort actually knew Jean Rodolphe makes him the most likely candidate. In addition to being a physician, he wrote a bibliography of the natural history of Switzerland and was distinguished enough to be elected to the Royal Society in Britain in 1708.

We can identify most botanical eponyms – from the Greek *eponymos*, meaning to give one's name to something or someone, from *epi* = upon + *onoma* = name – with more certainty, but even ones from the nineteenth century can be elusive. The giant American redwood, for example, was named *Sequoia* by Stephan Endlicher (1804–49), who in his short life published a Chinese grammar, a catalogue of Chinese and Japanese coins, and several works of botany, while also serving as director of the Vienna Botanical Garden. Endlicher seems never to have explained why he chose the name *Sequoia*, but it was assumed he was honouring Sequoyah (c. 1770–1843), a mixed-race Cherokee who had invented a written language for his people. The details of Sequoyah's life are equally contested, with some sources claiming that he was the son of a Swabian pedlar, others that his father was Scottish or himself of mixed race. What is certain, however, is that Sequoyah was born in what is now Tennessee, and the one extant species of the tree, *Sequoia sempervirens*, is a native of coastal California. An alternative suggestion is that Endlicher named his genus from the Latin word *sequi*, meaning to follow, because although the California redwood is the only species that survives, it is a 'follower' or remnant of several extinct species known about from fossils. This, too, has been rejected by taxonomists, and the most recent research at the time of writing states that Endlicher really did have his fellow-linguist in mind.

The further back we go in time, the less likely we are to recognise botanical eponyms, some of which commemorate people who are otherwise more or less forgotten except perhaps in the fields in which they worked. This is particularly true of genus names. If you know your history of chemistry, for example, you may know that the German scientist Christian Weigel (1748–1831) invented the laboratory cooling heat exchanger; but even those who grow weigela are unlikely to have heard of him. We all recognise a begonia when we see one, but (in Britain at any rate) few of us will have heard of Michel Bégon (1638–1710), the French colonial administrator and patron of a botanical expedition to the West Indies after whom the plant is named. And, further back still, what about Jean Nicot (1530–1600),

Tobacco, *Nicotiana*

the French ambassador to Portugal, who was introduced to tobacco in that country and sent seeds of this apparently miraculous medicinal plant back to Catherine de Medici in France? Although Nicot did not in fact discover tobacco, which had been introduced to Portugal from the Americas, Linnaeus named the plant's genus *Nicotiana* after him. When we see waves of purple aubrietia breaking over a stone wall, we ought really to think of Claude Aubriet (1665–1742), the brilliant botanical artist who provided illustrations for Tournefort's *Éléments de botanique* and accompanied the great man on his trip to the Levant in 1700. There are all sorts of reasons for remembering Aubriet's name, but an important one for gardeners is that it helps get right a plant name that is frequently misspelled.

That said, some plants commemorate people whose names are so complicated to spell that it is not much help to know them. Perhaps the most difficult of all is the species name of the so-called golden paeony, *Paeonia mlokosewitschii*. The peony in fact has flowers of the most delicate pale yellow and is named after Ludwik Młokosiewicz (1831–1909), a Polish officer who found this and several other plants while serving with the Russian army in the Caucasus. Given how reluctant the British are to pronounce foreign names – 'Molly the witch' being regarded as an acceptable stab at it – it is perhaps as well that this is the only plant that bears this specific epithet, although ornithologists have to struggle with another of Młokosiewicz's discoveries, the Caucasian black grouse, *Tetrao mlokosiewiczi*, spelled differently but no more easily. After various vicissitudes, including having his botanical collections confiscated and being sent into exile for six years when he was falsely charged with inciting Caucasian Poles to revolt, Młokosiewicz ended his life as an inspector of forests. He took his children on botanical excursions at a very young age, and at least one of them made her own contribution to botanical nomenclature by discovering a bright magenta primrose, to which she very sensibly gave her first name rather than her surname: *Primula juliae*. The other notorious plant name is that of the California poppy, *Eschscholzia californica*, named in honour of (if not in fact spelling correctly) Johann von Eschscholtz (1793–1831), a Baltic German doctor and naturalist who went in search of flora and fauna in Alaska, California and Hawaii. The correctly spelled Eschscholtz Atoll in the Marshall Islands was also named after him but, no doubt to the relief of all geographers, it was renamed the Bikini Atoll in 1946, when it became the site for several US nuclear tests.

Knowing how such surnames are pronounced isn't necessarily a reliable guide to pronouncing the botanical name derived from them, dahlia being a case in point. It may be technically correct to sound Anders Dahl's name when you speak of dahlias, but very few people – in Britain at any rate – do so, not least because it sounds ludicrously lah-di-dah. Some people pronounce choisya in proper recognition of the francophone Swiss botanist and philosopher for whom it is named, Jacques Denis Choisy (1799–1859), although the *OED* comes down firmly for the more usual 'choy-zia'. The *OED* gives alternative pronunciations (anglicised and Germanic) of *Deutzia*; Johan van der Deutz (1743–88), the Dutch lawyer who financed Thunberg on his trip to Japan, where he found the shrub in the 1770s, presumably pronounced his surname the latter way. Similarly, if you want to honour Johann Heinrich von Heucher (1677–1746), the German professor of medicine and botany after whom heucheras are named, then you would pronounce the plant 'hoykera', though the *OED* offers 'hewkera' as well, while Stearn adds 'hoyshera'.

Even when you have had a plant named after you that retains the correct sound and spelling, immortality is not guaranteed. Paolo Boccone (1633–1704), the Italian monk after whom the plume poppy was once named, slipped even further from view when 'his' genus was renamed *Macleaya*. Whether or not this helps you remember Alexander Macleay (1767–1848), the Scottish lepidopterist and Secretary of the Linnaean Society, is doubtful – unless you are Australian, since he became Colonial Secretary of New South Wales in 1825 and built a large mansion at Elizabeth Bay in Sydney that is now a museum, though the botanical garden he established there has been lost. Dr Olof Celsius, the dean of the cathedral in Uppsala and a keen naturalist whom Linnaeus had met in the city's Botanical Garden is another distinguished figure whose once famous name has been lost to botany. Impressed by the young student's knowledge of botany, Celsius had offered Linnaeus a room in his house and the run of his extensive library. In gratitude Linnaeus named a genus of mullein *Celsia* after him. Whereas the name of Olof's nephew Anders Celsius is still commemorated in the Celsius scale of temperature he invented, Olof's own immortality was overturned when the *Celsia* genus was subsumed into that of *Verbascum*.

The names of many plants are now so familiar to gardeners that they may not even realise they are simply Latinate versions of surnames. Bright yellow forsythia, for example, owes its name to William Forsyth (1737–1804), who trained under Philip Miller at the Chelsea Physic Garden (where the rockery he created in 1773 is now a Grade II listed structure), was appointed royal gardener to George III, and in the last year of his life was one of the seven founder members of the Horticultural Society of London, later the RHS. (Two other founder members had plants named after them: *Grevillea* for Charles Francis Greville, a Lord of the Admiralty, and *Dicksonia* for the Covent Garden nurseryman James Dickson.) And what about gardenias, magnolias and camellias? The first is named after Alexander Garden (1730–91), a Scottish-born doctor who emigrated to America, settling in Charleston and spending much of his time studying plants, specimens of which he sent to Linnaeus. The second commemorates the French botanist Pierre Magnol (1638–1715), who was director of the Montpellier Botanic Garden and the author of

Prodomus historiae generalis plantarum (1689), in which he laid out plants according to their families. The camellia is named after Georg Josef Kamel (1661–1706), a Moravian Jesuit lay brother who travelled to Luzon in the Philippines in 1688 to establish a free pharmacy for the local poor, and studied the island's flora and fauna, providing John Ray with an extensive appendix to his *Historia Plantarum*. Kamel clearly deserved honouring, but it is unclear why Linnaeus adapted his name for the *Camellia* genus.

Although people's names had provided botanists with plant names for several centuries, it was Linnaeus who turned the idea into a fine art, frequently saluting other botanists when creating genus and species names for his new system. As in the case of Kamel, his precise reasons for doing so are sometimes obscure. Why, for example, did he honour the Rev. Adam Buddle (1662–1715) by naming the *Buddleja* genus after him? Buddle was indeed a fine botanist, who compiled a complete British flora that, although unpublished, entered the collection of Sir Hans Sloane; but his speciality was mosses and grasses. It may also at first seem puzzling that Linnaeus named the zinnia after Johann Gottfried Zinn (1727–59), who was director of the Göttingen Botanical Garden, but his reason for choosing this particular plant may have been a little joke. As well as being a botanist, Zinn was also a distinguished anatomist who published the first study of the structure of the human eye, and it is possible that Linnaeus knew that in its native Mexico the zinnia is called *mal de ojos*, meaning 'the evil eye'.

In a neat reversal of the process of naming plants after people, Linnaeus's own name was derived from a plant. His father, Nils, who came from Småland, a southern province of Sweden, originally bore only a patronymic, Ingemarsson (son of Ingemar). On entering university, he was obliged to provide a surname and chose Linnaeus, a Latinised form of *lin*, the Smålandic word for a lime tree, because a large and celebrated one grew on his family estate. Our Linnaeus had only one plant named after him, a surprisingly unshowy but very pretty little woodland one native to Lapland. Originally identified by Gaspard Bauhin as *Campanula serpyllifolia*, it was renamed *Linnaea borealis* by Johan Gronovius, a Dutch botanist who, with the Scottish physician Isaac Lawson, financed the publication of Linnaeus's *Systema Naturae*. (The three men had met in Leiden, where Linnaeus, clutching sheaves of manuscripts, had gone to meet the leading doctor and botanist Herman Boerhaave.) Linnaeus subsequently adopted the flower as his emblem, and it features in many portraits of him. In his *Critica Botanica* (1737), which laid out rules for the naming of plants, he noted: '*Linnaea* was named by the celebrated Gronovius and is a plant of Lapland, lowly, insignificant, disregarded, flowering but for a brief space – from Linnaeus who resembles it.' Linnaeus was only twenty-nine when he published this book, but he already had a very high opinion of his own merits and the affected modesty of this description is at odds with the book itself in which he rejected a huge quantity of names created by his predecessors, replacing them with those of his own invention. This did not meet with universal approval, and he received a stern ticking off from Johann Amman, professor of botany at St Petersburg, who wrote: 'I beg you to consider what would happen if everyone were to lay down such laws and regulations whenever he felt so inclined, overturning names known and approved by the best authors just for the sake of making new ones.' Linnaeus did not take criticism kindly, and while Amman appears to have been forgiven, another critic from St Petersburg, Johann Siegesbeck, had the small and dreary weed *Siegesbeckia* named after him.

Linnaeus did, however, give the sponsors of *Systema Naturae* their due by naming the henna tree *Lawsonia* and another plant *Gronovia*, 'a climbing plant which grasps all other plants being called after a man who has few rivals as a "collector" of plants'. Neither of these plants is likely to feature in the average garden, but rudbeckias might. Olof Rudbeck the Younger was professor of medicine at the University of Uppsala, where Linnaeus went to study in 1728. By this stage in his career Rudbeck was spending more time pursuing his own studies than attending to students; he did, however, spot Linnaeus's undoubted talents and in 1730 appointed him a demonstrator in botany at the university. It was for this reason that Linnaeus named the coneflower *Rudbeckia* after him, writing in his best fawning style:

> I have chosen a *noble* plant in order to recall your merits and the services you have rendered, a *tall* one to give an idea of your stature; and I wanted it to be one which branched and which flowered and fruited freely to show that you are cultivated not only in the sciences but also the humanities. Its rayed flowers will bear witness that you shone among savants like the sun among the stars; its perennial roots will remind us that each year sees you live again through new works. Pride of our gardens, the *Rudbeckia* will be cultivated throughout Europe and in distant lands where your revered name must long have been known. Accept this plant, not for what it is but for what it will become when it bears your name.

Rudbeck was indeed distinguished, but would scarcely be remembered today if it were not for the popular garden plant that, as Linnaeus predicted, now grows in gardens all over the world. However, few of those who named plants to honour people went to quite the lengths Linnaeus did in explaining their reasons: in most cases it was merely a gesture.

Other plants that Linnaeus named after people he knew include the tender bush violet, *Browallia*, which he had grown from seed collected in Panama and given to him by Philip Miller. He named the plant after his erstwhile 'best friend' Johan Browallius, a clergyman who became Bishop of Åbo and subsequently fell out of favour. When Linnaeus visited London in the summer of 1736, Miller gave him a tour of the Apothecaries' Garden. The touchy Swede was displeased to find that Miller had not adopted his binomial system for this important garden, and when he renamed a negligible yellow member of the daisy family *Milleria*, it may not have been quite the compliment it seemed. By contrast, the English botanist Peter Collinson, who not only advocated the Linnaean system but proved a generous host to its inventor during the London visit, had the valuable medicinal plant *Collinsonia canadensis* named after him. Some of those favoured students whom Linnaeus unblushingly called his 'apostles' were botanically rewarded too: *Kalmia* was named after Pehr Kalm (1716–79), who had brought this evergreen shrub back to Sweden from his travels in North America; *Osbeckia* after Pehr Osbeck (1723–1805), who had collected some 600 specimens for Linnaeus in Canton, including the magenta-flowered *O. chinensis*; *Alstroemeria* after Clas Alströmer (1736–94), who sent seeds of the plant from Spain; and *Sparrmannia*, a large shrub sometimes called the African linden, after Anders Sparrman (1748–1820), who travelled to China and the Cape, and whose extraordinary life is told in Per Wästberg's wonderful 'biographical novel', *The Journeys of Anders Sparrman* (2010).

In South Africa Sparrman was joined by another of

Linnaeus's pupils, Carl Peter Thunberg (1743–1828), who was employed by the Dutch East India Company and on his way to Japan. Thunberg spent three years on the Cape, making a study of its people, plants and animals, and undertook several small expeditions with Sparrman before venturing further afield. He subsequently published a lively account of his travels as well as a *Flora Capensis* (1807). The genus of African plants that numbers the black-eyed Susan vine among its species was named *Thunbergia* after him, though not by Linnaeus. He eventually reached Japan in 1775 and began exploring the country, its people and its flora and fauna as far as the Japanese, always wary of foreign visitors, allowed. He spent some fifteen months there before returning to Sweden, via Java and Ceylon, and published his *Flora Japonica* in 1784. 'I am most anxious to live until you get back,' Linnaeus had written to Thunberg in 1773, when the latter was setting out on a long journey in Africa; 'what a joy it would be for me to be present on that great day, and to touch with my hands the laurels that will crown your brow.' These laurels did not include any plants named by Linnaeus after Thunberg, nor did the great teacher live to see the return of his pupil, who would eventually succeed him as professor of botany at Uppsala University. Other botanists, however, honoured Thunberg with the epithet *thunbergii* for species of berberis, pine, astilbe, geranium, allium, fritillary, hemerocallis, and spiraea (Thunberg's meadowsweet).

Another of Linnaeus's apostles, Daniel Solander (1733–82), emigrated to England, sailed on Captain Cook's *Endeavour* with the botanist Joseph Banks, and ended up as Keeper of the British Museum's Natural History Department. He gave his name to the Solander box or case, invented for the safe storage of plant specimens, manuscripts and other fragile objects and still widely used in libraries and museums. The trumpet-flowered *Solandra* is also named after him, and the specific epithet *solandri* identifies a number of New Zealand natives, including grasses (*Carex* and *Elymus*), the Coastal shrub daisy *Olearia solandri*, and the black beech *Nothofagus solandri*. Linnaeus's son, who had been taught by Solander and followed in his father's work, named a genus of protea *Banksia* in honour of his old teacher's friend and colleague, with whom Solander had collected the plant in Australia. The species *Banksia solandri* commemorates the friends together in one plant.

Linnaeus *père* acknowledged some of the earlier writer-botanists we have already come across, naming the lobelia after Matthias de l'Obel, and *Bauhinia* after the Bauhin brothers – for the rather prosaic reason that the leaves of this colourful tropical plant have two lobes. Given what a pioneer in identifying plants l'Obel was, it seems strange that he has so few of them named after him: the dwarf prostrate broom *Genista lobelii*, Lobel's maple *Acer cappadocicum* subsp. *lobelii*, and an unspectacular rotund cactus with nugatory pink flowers, *Melocactus lobelii*. Rembert Dodoens also deserved better than the rather dull genus of evergreen shrubs, *Dodonaea*, that Philip Miller named after him. The best that can be said for the genus is that the papery-winged fruit of *D. viscosa* was once used in Australia as a substitute for hops to give beer its distinctive bitter taste. The *Caesalpinia* genus is at least one containing many showy tropical and subtropical trees and shrubs, including Barbados pride, *C. pulcherrima*, but that is all Cesalpino gets, and no specific epithet, in spite of his important contribution to the naming of plants. It equally says much for the fate of Theophrastus that, apart from the Cretan date palm, *Phoenix theophrasti*, and Chinese jute, *Abutilon theophrasti*, his name is unacknowledged in plant taxonomy.

There was once a family of tropical trees and shrubs called the *Theophrastaceae*, which contained ninety-five species, including *Theophrasta americana* and *T. jussieui*, identified and named by the nineteenth-century botanist John Lindley. Posterity has treated Lindley better than Theophrastus: the RHS's Library in London carries Lindley's name, while the *Theophrastaceae* have been consigned to the overflowing taxonomical dustbin.

That other early writer on plants Dioscorides had the large but unshowy *Dioscorea* genus, which includes both yams and black bryony, named in his honour, but did rather better with the mottled purple-brown *Arum dioscoridis*, first described in John Sibthorp, James Edward Smith and Ferdinand Bauer's spectacular ten-volume *Flora Graeca* (1806–40), and with one of the best species of acanthus, *A. dioscoridis*, which has bright pink flowers. Stylised spiny acanthus leaves were widely used by the ancient Greeks in decorative devices, notably on the capitals of Corinthian columns, so it seems appropriate that the great Greek physician should have one of this species named after him. (The Roman form of the Corinthian capital tends to use the more rounded leaf form of *A. mollis*.) Linnaeus named the vast tribe of *Euphorbia* after another Greek doctor, Euphorbus, physician to Juba II, a ruler of Numidia and Mauretania in North Africa in the early decades of the Christian Era who married the daughter of Antony and Cleopatra. A legend that Euphorbus cured the king of a stomach ailment by administering the plant that eventually took his name seems at first unlikely, since the sap of most species is highly toxic. However, the English name of the plant, spurge, derives (via Old French) from the Latin *expurgare*, meaning to clean, and the plant was used as a purgative during the medieval period. Quite what the recovery rate was from this extreme remedy is not recorded, and one very good reason for knowing

'Sweet wood spurge', *Euphorbia*

the Latin names for plants is that it will prevent you from confusing the caper spurge, *E. lathyris*, with the caper bush, *Capparis spinosa*. The former takes its English name from the fact that its fruits resemble capers, but the entire plant is poisonous, whereas the flower buds of the caper bush are pickled and widely used in cooking. In fact the whole of the latter plant, excluding the roots, can be eaten. I somewhat gingerly tried this once at a restaurant in a remote Maronite village in Northern Cyprus, all too aware of the reason for its specific epithet, *spinosa*. The relationship between Euphorbus and Juba is commemorated by *Euphorbia regis-jubae*, a shrubby species of the plant native to the Canary Islands.

The *Eupatorium* genus mentioned in the first chapter owes its name to another ancient ruler, Mithridates Eupator, King of Pontus (135–63 BC). Opera lovers will recall that Mozart wrote an early opera about this hapless monarch, *Mitridate, re di Ponto* (1770), and poetry lovers will perhaps remember Mithridates from the penultimate poem in A. E. Housman's hardy perennial *A Shropshire Lad*. Housman was not only the leading classicist of his day, but a keen amateur botanist who would have had no trouble identifying the water-loving hemp agrimony, *Eupatorium cannabinum*, had he come across it growing in the western brooklands of his Worcestershire childhood. Having suspected his mother of attempting to poison him because she favoured his younger brother as heir to the throne, Mithridates started taking small doses of poison in order to provide homeopathic protection. He was also credited with concocting a herbal antidote to poison, which became known as mithridate and was highly valued during the medieval and renaissance periods. The medicine was reputed to contain some thirty-six ingredients, including saffron, cassia, rose petals and rhubarb root, and all manner of plants, from the costly frankincense (*Boswellia*) to such lowly weeds as darnel (*Lolium temulentum*) and shepherd's purse (*Capsella bursa-pastoris*). None of this did Mithridates much good: captured by his principal enemy, the Romans, he was unable to use poison to kill himself in captivity, an irony Housman particularly relishes in his poem. He did not have a great deal of luck in botany either, since although hemp agrimony is still in the *Eupatorium* genus, garden species such as *Ageratina altissima*, the one I mentioned growing in my own garden, and purple Joe-Pye weed, *Eutrochium purpureum*, have been shifted to other genera. One final plant named for a ruler from the ancient world is *Lysimachia*. According to Pliny, this genus honours Lysimachus, who became king of Thrace in 306 BC. The ruler's name means 'ending strife', hence the English name of the plant, yellow loosestrife – though Geoffrey Grigson records that the origins of the name may be more lowly because the plant was reputed 'to end strife between horses and oxen yoked to the same plough'.

Because botanical names derive from Latin and Greek, it is unsurprising that a good number of them look back to the ancient world and to figures from classical mythology. Yarrow takes its botanical name, *Achillea*, from the Greek hero Achilles, whose education was entrusted by his parents to the centaur Chiron, after whom both the *Centaurium* and *Centaurea* genera are named. Centaury was widely thought to be particularly effective in healing wounds, and Culpeper recorded that knapweeds were used to treat 'wounds and ruptures, bruises, sores, scabs, and sore throat, etc'. Medicine was clearly an important part of the curriculum at Chiron's academy, and it was here that Achilles was supposed to have learned about the healing properties of yarrow (sometimes called soldier's woundwort) – not that it did him much good when he died after being hit by an arrow in the one vulnerable part of his body, the heel, during the Trojan War.

Another hero of that war was Ajax, who had been educated alongside Achilles by Chiron. After Achilles' death, Ajax and Odysseus retrieved his body and then quarrelled over who should inherit the armour. Odysseus won, and Ajax fell on his own sword. Despite being such a bad loser, Ajax had *Delphinium ajacis* named after him. Depending on who you believe, the larkspur either sprang up from his blood or acquired its species name because its petals were marked with the Greek letters *AI AI*, meaning 'Alas! Alas!' The Trojan War had been fought over Helen, the wife of King Menelaus of the Spartans, who had eloped with the Trojan prince Paris and who gave her name to *Helenium*. Quite why heleniums, attractive as they are, are named after someone reputed to be the most beautiful woman in the world is a puzzle. There was, however, a myth that another plant of the same family, elecampane (*Inula helenium*), sprang from the tears Helen had shed. Elecampane is a large, rather coarse daisy, but then Helen had a great deal to cry about. The plant's association with Helen led to elecampane being used by women in the ancient world to enhance their beauty, either in a face-pack or in a decoction. According to Pliny, 'women think that they acquire a kind of aura of attractiveness and sex appeal by its use'. This use is not mentioned by Gerard, though he does suggest an ointment made from the plant 'doth clense and heale up old ulcers'. Gerard gives elecampane the genus name of *Helenium*, and further slippage seems to have led to this being transferred to the plant we now know by this name.

Other flowers were reputed to have sprung from the spilled blood of figures from the ancient world. The bright scarlet pheasant's eye, *Adonis aestivalis*, was said to have appeared after Aphrodite's mortal lover Adonis was fatally gored while hunting wild boar on Mount Lebanon. The Abraham or Adonis River, which flows down from the mountain, is stained red by the clay that washes into it after the winter rains. In ancient times it was believed that the water was still being coloured by the blood of Adonis, around whom a cult built up at Byblos, where the river flowed into the sea. By some curious linguistic cross-fertilisation, the anemone may also get its botanical name from Aphrodite's slain lover. Anemones were traditionally known as windflowers because it was thought the name derived from *anemos*, the Greek word for wind: for, as William Turner stated, 'the floure never openeth it selfe, but when the wynde bloweth'. Another suggestion is that the name derives from 'Naaman'. According to *The Golden Bough*, J. G. Frazer's classic 1890 study of comparative religions, Naaman means 'darling' and was an epithet for Adonis in Phoenician rituals commemorating his death; the Arabic word for the flower, *shaqa'iq An-Nu'man*, translates as 'wounds of the Naaman'. When seen growing wild *en masse* – in Israel, for example, where it has been adopted as the national flower – the original scarlet form of *Anemone coronaria* has a similar visual impact to the bloody red Flanders poppy, *Papaver rhoeas*, which was such a feature of the Western Front in the First World War that it became a symbol of those who bled and died there. Aphrodite, incidentally, lends her name to the genus of the lady's slipper orchid. One of the proposed places of her birth was the island of Cyprus, which gave her an alternative name of Cypris, and *pedilon* is the Greek word for a slipper, hence *Cypripedium*.

It was Adonis's mother, Myrrha, sometimes known as Smyrna, who lent her name to the genus *Smyrnium*, which includes among its species the impressive but invasive umbel-lifer S. *olusatrum* (alexanders),

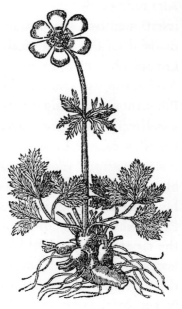

Scarlet anemone,
Anemone coronaria

and that favourite of fashionable flower-arrangers, the fresh-green and lime-yellow biennial S. *perfoliatum*. As Ovid warns when introducing her story in his *Metamorphoses*, Myrrha was badly behaved even by the standards of classical mythol-ogy and harboured a sexual passion for her own father, King Cynaris of Cyprus. Adonis was conceived after Myrrha had tricked Cynaris into bed. Discovering that the young woman with whom he had been happily sleeping in total darkness was in fact his own daughter, Cynaris attempted to kill her. Myrrha escaped into exile and spent long months wandering round the Arabian peninsula until the gods took pity on her and trans-formed her into a myrrh tree of the *Commiphora* genus. This name means 'resin-bearing' and myrrh is the fragrant resin these trees exude, in mythology representing Myrrha's tears.

Alexanders takes its common name from Alexandria, though it is in fact native to the Canaries. The plant was, however, widely distributed around the Mediterranean and the Romans, who imported it to Britain as a food crop, called it the parsley of Alexandria. Its botanical genus name came about because Pliny thought it smelled similar to myrrh; the species name was created from *olus* = vegetable + *ater* = black, because of the very distinctive black seeds.

The hyacinth was supposed to have sprung from the spilled blood of another beautiful youth who was loved by gods. Hyacinthus was a Spartan prince who attracted and reciprocated the amorous attentions of Apollo, but had also caught the fancy of Zephyrus, the god of the west wind. One day, hoping to impress his lover, Hyacinthus attempted to catch a discus that Apollo had thrown and was struck dead. In other versions of the story the jealous Zephyrus deliberately blew the discus off course in order to punish Apollo by killing his boyfriend. It seems, however, that the plant that sprang up from his blood – the petals of which were marked by Apollo's tears with the same Greek letters, *AI AI*, that supposedly appear on Ajax's larkspur – was almost certainly not the hyacinth we know today. Indeed, it may even be the same plant as Ajax's, a mythical story shared. That said, Linnaeus named the English bluebell *Hyacinthus non-scriptus*, meaning 'not written upon', in order to differentiate it from Apollo's tear-stained one. The bluebell has been renamed several times since then and although it is now *Hyacinthoides non-scripta*, for a while it was named *Endymion non-scriptus* after another beautiful youth, the shepherd-prince with whom the moon goddess Selene fell in love and who is the subject of Keats's long narrative poem of 1818.

Greek gods being what they were, when not sporting with

beautiful youths, Apollo spent time pursuing goddesses and nymphs, including a naiad or water sprite called Daphne. Less susceptible to Apollo's suit than Hyacinthus, Daphne pleaded with her father, a river god called Peneus, to rescue her. He did so by transforming her into a laurel, a subject much loved by painters and particularly sculptors because the moment of metamorphosis was a test of their skills. The Greek word for laurel is *daphne*, and this name became transferred to the familiar shrub with its highly scented flowers. Naiads also provided a name for the water-lily genus, *Nymphaea*, and in Alfred Waterhouse's famous 1896 painting *Hylas and the Nymphs*, naked naiads are seen surfacing in a pond covered in water-lilies in order to lure the young prince to his doom. It was the Nereides, or sea-nymphs, who gave their name to the *Nerine* genus of autumn-flowering bulbs with bubblegum-pink flowers. The plant is native to South Africa, but the original scarlet-flowered species takes its name from Guernsey and is *Nerine sarniensis*, Sarnia being an old name for that Channel Island. There is a legend that bulbs of this plant were washed up on the island's beaches after a shipwreck and naturalised there, which is why the genus was named after sea-nymphs and the plant came to be known as the Guernsey lily; but the bulbs were in fact rather more prosaically presented to John de Saumerez, first Dean of Guernsey, by sailors whose ship had run aground in the mid-seventeenth century. One of the sea-nymphs was Pleione, who as well as being the mother of the Pleiades, the seven sisters who became a familiar cluster of stars in the night sky, gave her name to a genus of orchids.

The name of the Greek god of medicine, Asclepios, was borrowed by Linnaeus for a genus of milkweed, *Asclepias*. Given that most milkweeds are highly toxic, it seems a strange choice – though *A. tuberosa*, which has bright orange flowers

'The double white Peionie', *Paeonia*

that attract butterflies, was widely used to treat pulmonary problems and known as pleurisy root. The peony also owes its name to an ancient physician, Paeon, who appears in Book V of the *Iliad* ministering to the wounded gods: 'Hades went to the house of Jove on great Olympus, angry and full of pain; and the arrow in his brawny shoulder caused him great anguish till Paeon healed him by spreading soothing herbs on the wound.' Paeon later performs a similar service for Mars: 'As the juice of the fig-tree curdles milk, and thickens it in a moment though it is liquid, even so instantly did Paeon cure fierce Mars.' Peony seeds taken in wine were recommended by Dioscorides for uterine problems, a cure that was still being followed over 1500 years later when Parkinson thought it efficacious for 'the rising of the mother' or womb. Parkinson also recommended the root of the 'Male Peonie' as 'a most singular approved remedy for all Epilepticall diseases', and parents often strung these roots around the necks of children to ward off 'the falling sicknesse'. According to Gerard, peonies could also be used for jaundice, kidney and bladder pains, cleansing the liver and preventing nightmares, so it is perhaps unsurprising the plant should be named after a famously successful ancient physician.

One plant too dangerous to be in the pharmacopoeia was deadly nightshade, which also takes its botanical name from

Greek mythology. Gerard records that the plant's leaves 'may with great advice be used in such cases as Petti-mortell', which is presumably some kind of serious fainting fit, but adds: 'if you will follow my counsel, deale not with the same in any case, and banish it from your gardens and the use of it also, being a plant so furious and deadly'. By way of illustration, Gerard recounts a story about three boys from Wisbech in the Cambridge Fens who were tempted by the plant's beautiful shining black berries: within eight hours of eating them, two of the children had died, while a third was saved only by being repeatedly dosed with a honey emetic. It was the plant's grim reputation that gained it the botanical name *Atropa bella-donna*. Atropos was one of the three Fates who controlled the lives of men, and it was she who wielded the shears to cut through the thread of life. The specific epithet is taken from the plant's sixteenth-century Venetian name of *Herba bella donna*, recorded by Matthiolus, who reported that the city's women used a distillation of it to enlarge the pupils of their eyes and so enhance their beauty.

Returning from renaissance Venice to the ancient world we find a family relationship between liriope – most familiar in gardens as the species *Liriope muscari*, which has striking lilac flower-spikes – and the daffodil. Liriope was a water-sprite from the kingdom of Boeotia, where the great city of Thebes once stood. She caught the eye of the river god Cephisus, who 'clasped [her] in his winding streams, and took [her] by force under the waves'. Narcissus was the result. The story of Narcissus, who fell in love with his own reflection after seeing it in a pool, has been told many times, but it is in Ovid's *Metamorphoses* that the beautiful youth turns into a flower. After he drowned, his sisters prepared a funeral pyre, but he meanwhile vanished and 'they came upon a flower, instead of his body, with white petals surrounding a yellow heart'. Some varieties of narcissus have drooping heads when they are in bud, and this mourning posture is why they are associated with their eponym's death. The highly scented *Narcissus jonquilla* (so called because of its narrow foliage, from the Latin *juncus* = reed) is in fact a native of the Iberian peninsula rather than of Greece, but 'jonquil' became a generic poetic term for the narcissus, of which there are some fifty species and innumerable cultivars, some of them, alas, far removed in appearance from the elegant simplicity of the flower Ovid described. There is some dispute as to whether *Narcissus jonquilla*, the bunch-headed *N. tazetta* (of which the cultivar 'Paperwhite' is particularly popular), or the unsurpassable 'old pheasant's eye', *N. poeticus*, was the first daffodil to be cultivated, but the last named is mentioned by Theophrastus as one of the earli-

est flowers to appear. This may have been the case in ancient Greece, but in the English climate it is one of the later-flowering daffodils, usually in bloom between March and May. Narcissi are also among the flowers with which the love-struck shepherd Corydon attempts to seduce the beautiful boy Alexis in Virgil's fifth Eclogue, which was published around 39–38 BC.

Helios, the Titan who rode his chariot across the sky, was a personification of the sun and so naturally gives his name

Jonquil, *Narcissus jonquilla*

to several plants. *Helianthus* combines the Greek words for sun and flower, *helios* + *anthos*, although in Gerard's day the common sunflower, *H. annuus*, derived its botanical name from Latin rather than Greek, and was known as *Flos solis*, reputedly 'taking that name from those that have reported it to turne to the Sun, the which I could never observe'. Gerard thought instead that the plant got its name because the huge individual flower-head 'doth resemble the radiant beames of the Sun'. Indeed the flowers look exactly like stylised images of the sun, with the bright yellow petals as the rays surrounding the flaming orange centre. Unsurprisingly the plant has been widely used as a decorative device and became a symbol of the late-nineteenth-century Aesthetic Movement in Britain, frequently accompanying Oscar Wilde in caricatures, flourishing in the books of Kate Greenaway, and moulded upon terracotta tiles on buildings. The unrelated but sun-loving rock rose, *Helianthemum*, derives its name from the same source and is sometimes called the sun rose – though it is not related to the rose and in fact belongs to the cistus family. Before hybridisation dramatically increased the colour range, helianthemums mostly had sun-yellow flowers, 'small like little Roses', as Gerard observed. Heliotropes have flowers that are purple rather than yellow, but like those of helianthus, they were once believed to follow the sun – *Heliotropium* from *helios* + *trepein* = to turn – and their old English name was Tornesole. Gerard dismissed this idea, writing instead that this name was adopted because the plant 'flowereth in the sommer solstice, at which time the sun being farthest from the Aequinoctiall circle, returneth to the same'. Tortuous as this sounds, heliotropes do indeed flower in June. Other plants with similar names include *Helichrysum* (*chrysos* = gold), which burns yellow and orange and doesn't fade even when picked and dried. (The *Concise*

Oxford English Dictionary's assertion that the first part of the name derives not from *helios* but from *helix*, meaning a spiral, is surely wrong?) The resemblance of the yellow daisy *Heliopsis* to the sun is similarly embodied in its botanical name, *opsis* meaning 'appearance'. On the other hand, *Heliconia* has nothing to do with the sun, but is named after Mount Helicon. Early taxonomists occasionally got carried away by their classical learning, and these tropical plants do not of course grow on the Greek mountain; but this was where the Muses lived, and it was thought heliconias were related to bananas, for which the genus name is *Musa*. The fact that *Musa* has nothing at all to do with Calliope, Clio, Terpsichore and their sisters, but is derived either from the Arabic word for the fruit, *Mauz*, or (less likely, and why?) from the Emperor Augustus's doctor, Antonius Musa, seems not to have deterred taxonomists from this kind of obscure punning. (Musa, incidentally, was the brother of Euphorbus.)

Alongside Helios in the skies of the ancient world was Iris, the messenger of the gods who also personified the rainbow – she arrives on stage perched on one in Handel's opera *Semele*. The name was given to the plant because of the variety and clarity of colours found in its petals, although Theophrastus barely mentions these at all, being more concerned with the rhizome, used to make perfume and to assist in childbirth. Few

'Turky Floure-de-luce', *Iris*

midwives these days will be found using the plant, but orris root – the rhizomes of the many-coloured bearded iris *I. germanica* and the generally lilac sweet iris *I. pallida* – provide an oil still used in the manufacture of modern perfumes, 'orris' being a sixteenth-century corruption of 'iris'. Iris sometimes carried messages for Zeus, and the pink was named as her employer's own flower, *Dianthus* derived from the Greek *Dios* = of Zeus + *anthos*. Zeus's daughter Hebe, who spent much of her time handing round ambrosia and nectar on Mount Olympus, lends her name to the popular genus of evergreen shrubs.

Within a genus named after one person, individual species names are sometimes drawn from a range of other people across many centuries. While *Euphorbia* commemorates a doctor from around the time of the birth of Christ, the giant *E. characias* subsp. *wulfenii* was named after Franz Xaver von Wulfen, a Jesuit and botanist born in Belgrade in 1728. Several other species of euphorbia owe their names to people who discovered them or introduced them into cultivation. *E. × martini* is named after the botanist Bernardin-Antoine Martin, who found this invasive but indispensable natural cross between *E. amygdaloides* and *E. characias* growing in the Gard, the region of France for which he published a flora in 1893. The ferociously thorny *E. milii*, with its startling red bracts, is named after Pierre Bernard Milius (1773–1829), the French governor successively of Île Bourbon (now Réunion) and Guyana, who introduced this Madagascan species to France. *Euphorbia griffithii*, notable for the contrast between its dark foliage and fiery orange bracts, was named by William Hooker, director of Kew Gardens in the mid-Victorian period, for William Griffith (1810–45), a bumptious young army doctor who collected plants in both Bhutan, where this species is native, and in India. Griffith also had *Ceratostigma griffithii* (Griffith's plumbago) and an arisaema named after him, and his knowledge of plants was such that in 1842 he was called to stand in for Nathaniel Wallich both as superintendent of the East India Company's Botanic Garden in Calcutta and as professor of botany at the city's medical college while Wallich was on a two-year leave of absence in South Africa. Griffith's great ambition was to compile a flora of South Asia, but having returned from Calcutta to Malacca, where he held the post of Civil Surgeon, he died of a liver disease aged only thirty-four.

The most charming story about the naming of euphorbias is that of *E. amygdaloides* subsp. *robbiae*, named after Mary Anne Robb (1829–1912). Widowed after just two years of marriage, Mrs Robb bought a large house at Liphook in Hampshire and gradually set about creating a garden there. She donated seeds and plants to the Royal Botanic Gardens at Kew, and when travelling in Greece arranged to be taught about indigenous plants by Theodor von Heldreich, a former director of the Royal Garden in Athens, who himself had many plants named after him, including the Greek maple, a deep pink allium, and the huge conifer *Pinus heldreichii*, introduced to cultivation from specimens collected on Mount Olympus. Mrs Robb also visited Turkey and supposedly brought back in her hatbox specimens of a euphorbia she had discovered there, which was subsequently named after her. This method of transportation led to the plant acquiring the colloquial name of Mrs Robb's bonnet.

A 'Biographical Memoir' of William Griffith published in the *Madras Journal of Literature and Science* noted that the young doctor often had to write both medical and botanical reports 'whilst labouring under illness, the effect of fatigue and exposure', and that 'his contempt of danger frequently brought him into hazardous collision with the people of the uncivilised tracts which he was engaged in exploring'. Many of those who travelled to distant lands in search of the plants we now buy with barely a thought for their origins had similar or worse experiences. And not all of them had their names botanically immortalised. One thinks of John Banister, the young cler- gyman sent to the Americas by the Bishop of London in the seventeenth century, who appears either to have fallen from a cliff while attempting to prise a plant from a crevice or been accidentally shot by a fellow explorer; or the hapless Charles de la Condamine (1701–74), who spent eight months ferrying a collection of young plants 2,500 miles down the Amazon, only to have them washed overboard when he reached the Atlantic. Joseph de Jussieu (1704–79), who introduced the common heliotrope to Europe from Peru, spent fifteen years studying plants in South America, at the end of which his col- lections were stolen and he went mad. To add insult to injury the genus *Jussieua* and the specific epithet *jussieui* refer not to him but to his more famous and successful brother Bernard, who was professor of botany at the Jardin des Plantes in Paris and never travelled further than Normandy. Francis Masson (1741–1805), who hunted plants in South Africa and the Caribbean for the Royal Botanic Gardens at Kew and brought back innumerable finds, including pelargoniums, gladioli and agapanthuses, found himself conscripted into the militia and taken prisoner of war when the French attacked Grenada in 1779. He was later captured by pirates during a trip to North America and threatened with execution, but was eventually handed over to another vessel and continued his journey, spending several years collecting specimens in both the United States and Canada. He did at least have a genus, *Massonia*, named after him, albeit a somewhat obscure if decorative ground-hugging South African member of the asparagus family. Dr Clarke Abel (1780–1826) arrived in China in 1816 as physician to a diplomatic mission. He was accompanied by an assistant supplied by Sir Joseph Banks, who was now advisor to George III at Kew, and collected a great many plants and seeds. On the voyage back to England his ship ran aground on a reef in the Indonesian Straits of Banca and everyone took to the boats, carrying what they could. Abel had managed to save a collection of seeds, but this was thrown overboard to make way for chests containing the linen of another, grander passenger. Abel returned to the wrecked ship the following day and managed to load more of his specimens on to a raft, which was promptly fired upon by pirates and completely destroyed. His party eventually found another ship to get them back to England, but all that Abel had left of his botanical treasures were some duplicate specimens – fortunately including the plant to which he gave his name, *Abelia chinensis*.

Perhaps the saddest story of plant collecting is that of the Scottish botanist David Douglas (1799–1834), after whom the Douglas fir is named. Douglas was to discover that the salary he received to slog across North America in appalling conditions was half what his employers at the Horticultural Society of London were paying their hall porter. Rendered virtually blind by years spent trekking through the snow and across open, hot, sand-strewn areas in the Columbia River basin, he eventually tumbled into a wild-cattle pit in Hawaii and was gored or trampled to death by a bull that was already in residence. Douglas was buried in a common and unmarked grave, but is commemorated by the specific epithets *douglasii* and *douglasianus*. These have been used for species small and large, from the poached-egg plant, *Limnanthes douglasii* to the blue oak, *Quercus douglasii*. While the California mugwort,

Artemisia douglasiana, is barely garden-worthy, the lavender-flowered *Iris douglasiana,* which Douglas found growing wild on the Pacific Coast, has won an RHS Award of Garden Merit – as well it might, given the Society's treatment of him. Among the many other plants that bear Douglas's name are the small mat-forming Columbia or tufted phlox, *Phlox douglasii;* the dangerously invasive *Spiraea douglasii,* flatteringly known as the rose spiraea because of its grubby pink flowers; and the black hawthorn, *Crataegus douglasii.*

Oddly enough, the Douglas fir itself bears the botanical name of *Pseudotsuga menziesii* after Douglas's fellow Scot Archibald Menzies (1754–1842), who first described it in 1791. Menzies too made expeditions to North America and had his own fair share of what Douglas called 'the not infrequent disasters attending such undertakings'. He may have had the good fortune to be served the seeds of the Chile pine, *Araucaria auraucana,* for pudding while dining with the country's governor in Santiago, which led to the introduction to Britain of the hugely popular monkey puzzle tree, but he argued with the captain of the ship on which he was sailing home from the Pacific Northwest and was confined to his cabin while the ship's chickens scratched up painstakingly gathered specimens of other plants. The *Menziesia* genus of flowering shrubs was named in his honour, but in 2011 it was reallocated to the *Rhododendron* genus, though *Menziesia ferruginea* has at least been renamed *Rhododendron menziesii.* This specific epithet has also been retained (so far) for a number of other plants. Among trees are the North American madrona tree *Arbutus menziesii;* the Australian firewood banksia, *Banksia menziesii;* and the New Zealand beech *Nothofagus menziesii,* still widely known by this name although it became *Lophozonia menziesii* in 2013. Among smaller plants are the pretty little *Sanguisorba menziesii* with slender maroon flower-spikes like miniature bottlebrushes; the delicate but rampant *Tolmiea menziesii*, the tubular purple and green flowers of which reward close inspection; and *Nemophila menziesii*, the common name of which, baby-blue-eyes, is an argument in itself for Latin binomials. Lest Menzies's ghost grow too complacent, the California blackberry, *Rubus menziesii*, was subsequently renamed *Rubus ursinus* (see 'Bestiary').

Other British plant hunters of the same generation include William Lobb (1809–64), whose name was most famously given to the 'Old Moss' rose. He was sent to Chile by the prominent nurserymen Messrs James Veitch & Sons to collect araucaria seeds on a commercial scale, which he did by blasting the cones out of the trees with rifles. Lobb was also responsible for introducing that other favourite of large Victorian gardens the Wellingtonia (*Sequoiadendron giganteum*), the holly-like *Desfontainia spinosa*, the yellow-flowered *Berberis darwinii*, the useful hedging plant *Escallonia rubra* var. *macrantha* and the Chilean fire bush, *Embothrium coccineum*. However, a large consignment of rare plants he had entrusted to a shipping agent in Ecuador never reached Britain but lay forgotten in a warehouse and was mostly destroyed by ants. When he failed to return from a plant-hunting trip to California after his contract with Veitch came to an end in 1857, it was thought that he might have transferred his attention to gold prospecting, although for a while he continued to send collections of seeds to England. Gradually, however, he ceased to correspond with friends and family. In 1863 he was 'seized with paralysis [and] lost the use of his limbs', possibly as the result of syphilis, and shortly afterwards died in a San Francisco hospital.

Veitch's Nursery was founded at Killerton in Devon at the beginning of the nineteenth century, and moved to new premises in Exeter in 1832. In 1853 the company acquired a second site when it took over the Royal Exotic Nursery in the King's Road, Chelsea, where Messrs Knight & Perry had conducted a very successful business. James Veitch was keen not only to sell known plants but to discover new ones and his nursery would eventually be responsible for the introduction of over 1,200 new species and cultivars. In 1906 it was acknowledged that *Curtis's Botanical Magazine*, founded in 1787 and famous for its lavish colour illustrations of plants, had relied upon specimens grown by the nursery for 422 of its plates. Alongside his brother Thomas, William Lobb had been employed by Veitch as a gardener at the Exeter nursery, and was first sent to South America in 1840. Some three years later Thomas too was dispatched abroad, first to Java, but subsequently travelling in other parts of Indonesia, the Philippines, Burma, north-east India and the Himalayas, where he discovered the *Phalaenopsis* genus of orchids, one of which, *P. lobbii*, was named after him. The genus *Lobbia* in the *Aristolochiaceae* or pipe-vine family was named when Thomas brought back a specimen from Singapore but was intended to honour both brothers; it is now more usually known as *Thottea*. The specific epithets *lobbii* and *lobbianum* honour the brothers as individuals, with William scoring rather better: a Californian poppy, *Eschscholzia lobbii*, the flowering currant *Ribes lobbii*, and an Ecuadorian sage, *Salvia lobbii*, are among the plants named after him. In addition to the orchid, Thomas is commemorated by a rhododendron he had collected in Burma with bright scarlet tubular flowers, *R. lobbii*. Thomas eventually retired after the amputation of one of his legs, which he had either injured or lost to exposure during a plant-collecting trip.

Some of the plants the brothers discovered were named after their sponsor, including *Rhododendron veitchianum*, a species with large fragrant white flowers that Thomas had found in Burma, and the insectivorous Veitch's pitcher-plant, *Nepenthes veitchii*, which he brought back from Borneo. This seems only just, since James Veitch & Sons sent many plant hunters to distant lands in search of new species. An obituary of James Veitch junior, the son of the nursery's founder, who died suddenly in his fifties in 1869, stated: 'Were we to attempt to show how far our gardens are indebted to the herculean and unflagging labours of Mr Veitch, we should have to write a history of most of the new plants introduced in the last thirty years; for it was to his active superintendence of their impor- tation, and to his discriminative choice of collectors, that we may largely attribute the success which was realised in this department.' The Veitch Memorial Medal was founded in his memory shortly after his death and is still awarded annually by the RHS to 'persons of any nationality who have made an outstanding contribution to the advancement and improve- ment of the science and practice of horticulture'. The nursery business was then taken over by James's two eldest sons, John and Harry. The former had undertaken plant-collecting trips to Japan, China and the Philippines, bringing back (amongst many other plants) the Japanese conifer *Abies veitchii* and such popular hothouse exotics as *Pandanus veitchii* and *Aralia vei- tchii*, but he died of tuberculosis aged only thirty-one. (He also introduced the umbrella pine, *Sciadopitys verticillata*, which we last saw when describing the death of Pliny.) Harry then ran the business alongside John's son James H. Veitch until the First World War, after which the Chelsea nursery closed, although the Exeter one continued trading, run by members of the family, until 1969.

Among other collectors employed by Veitch who suffered the frequent fate of such pioneers was David Bowman (1838–68), who sent back the widely grown houseplant *Dieffenbachia bowmanii* and several other plants before succumbing to dysentery in Bogotá; the young Henry Hutton, who in the 1860s discovered several orchids – *Cymbidium huttonii* and *Aerides huttonii* in Java and *Dendrobium huttonii* in Timor – before 'his health broke down and he succumbed to his enthusiasms'; and Gottlieb Zahn, who arrived in Panama in 1869, sent back a number of orchids and ferns before drowning the following year while on his way to Costa Rica, and had the air plant *Tillandsia zahnii* and the colourful bromeliad *Caraguata* (now *Guzmania*) *zahnii* named in his memory. J. Henry Chesterton was a former gentleman's valet who during the 1870s collected orchids for Veitch in South America, including the curled orchid *Odontoglossum crispum* var. *chestertonii* and the distinctly sinister dracula or frog's-skin orchid, *Dracula chestertonii*, but overestimated his powers of recuperation from a serious illness, insisted upon making a trip up river, and 'was barely put on shore at Puerto Berrio ere he died' in 1883. Gustave Wallis, a German collector who was deaf and dumb until the age of six, worked in South America, introducing the huge-leaved king anthurium, *Anthurium veitchii*, and a number of orchids before succumbing to a fever and dysentery in Panama in 1879. Wallis did, however, have a significant number of plants named after him – the orchids *Masdevallia wallisii* and *Epidendrum wallisii* and the popular indoor plants *Dieffenbachia wallisii*, *Curmeria wallisii* and *Maranta wallisii* – and was considered a good employee. Others failed to come up to the nurseries' exacting standards. The specific epithet *endresii* used for a number of orchids commemorates A. R. Endres, a mixed-race collector of these plants for Veitch in Guatemala and Costa Rica in the 1870s – and should not be confused with *endressii*, honouring the short-lived German

Philip Anton Christoph Endress (1806–31) and familiar from the bright pink (and sometimes misspelled) *Geranium endressii*. Veitch acknowledged that Endres sent many orchids back to England, 'but few of horticultural merit', while an expedition in 1873 was judged 'expensive and scarcely a success'. Even so, a couple of orchids Veitch presumably found saleable bear Endres's name, including the delicate Central American *Epidendrum endresii* and the insectivorous *Utricularia endresii*. Carl Kramer collected in both Costa Rica and Japan in the 1860s; despite introducing the orchid *Odontoglossum krameri* from the former country, he was considered by his employers to have proved 'entirely unsuitable for the work he had undertaken', lacking 'that adaptability and resource essential for successful exploration'.

This could certainly not be said of Veitch's most famous collector, Ernest Henry Wilson (1876–1930), who is credited with discovering over one thousand previously unknown species, and introducing innumerable plants, mostly from China. It was for this reason that he was popularly referred to as 'Chinese' Wilson. After leaving Veitch's employment, he collected plants for Charles Sprague Sargent (1841–1927), director of the Arnold Arboretum at Harvard University, extending his travels to Japan, Korea and Taiwan. Among his introductions to Britain were the handkerchief tree (*Davidia involucrata*), the beauty bush (*Kolkwitzia amabilis*, named for the German professor of botany and sewage-disposal expert Richard Kolkwitz, 1873–1956), the yellow lampshade poppy (*Meconopsis integrifolia*), the kiwi fruit (*Actinidia deliciosa*) and the highly scented *Lilium regale* and *Clematis armandii*, alongside quantities of rhododendrons, acers, abelias, cherries, viburnums, berberises and primulas. A hefty catalogue of 'woody plants collected in Western China' by Wilson was edited by Sargent and published between 1907 and 1910 in three volumes running to 1,966 pages, with the appropriately Latinate title *Plantae Wilsonianae*. Some sixty plant species bear his name, most notably the spectacular *Magnolia wilsonii*, with the pure white petals of its drooping flowers surrounding dark crimson stamens and carpels, and the vigorous and strongly scented *Clematis montana* var. *wilsonii*. The epithet *wilsonianus* also commemorates the great explorer, from the small blood-red rock-garden tulip *Tulipa wilsoniana* to the giant yellow-spiked bog-plant *Ligularia wilsoniana*. Wilson's nickname has also entered the botanical dictionary as a rare member of the witch-hazel family, *Sinowilsonia henryi*. The species name of this plant commemorates Augustine Henry (1857–1930), who collected specimens during his spare time while working for the Chinese Imperial Customs Service and had numerous species named in his honour, including *Acer henryi* and *Rhododendron augustinii* (so named because there was already a *R. henryi*, named after someone else), but is best remembered for the speckled orange *Lilium henryi*. E. H. Wilson's daughter Muriel is honoured by the umbrella bamboo *Fargesia murielae*, while his sponsor Charles Sargent is commemorated by a large number of trees and shrubs, including the pink-flowered *Magnolia sargentii*, the Japanese cherry *Prunus sargentii*, the Chinese rowan *Sorbus sargentiana*, and the pale yellow alpine *Rhododendron sargentianum*.

The handkerchief tree had been first described by the French missionary Armand David (1826–1900), hence the genus name of *Davidia*. While in China, David combined spreading the word of God with work as both a botanist and a zoologist, and he discovered many species unknown in the west, including the giant panda. Unusually, he gives both his forename and surname to a number of species, and his ecclesiastical title to a breed of Chinese deer, Père David's deer (*Elaphurus davidianus*). Two more of E. H. Wilson's introductions, the evergreen *Clematis armandii* and *Viburnum davidii*, which is grown as much for its metallic-blue fruits as for its clusters of small white flowers, were named in the missionary's honour. David also discovered *Rhododendron davidii* and *Photinia davidiana*, but the most familiar of the plants named for him, which he was the first person to record, is the butterfly bush *Buddleja davidii*, the lilac flowers of which have a distinctive orange eye. The snakebark maple, *Acer davidii*, was also discovered and recorded by David, but was introduced to Britain by another of Veitch's collectors in the Far East, Charles Maries (1851–1902). Maries had an eventful life: the plants and insects he had collected in Japan in 1877 were loaded on to a ship that was also transporting seaweed which swelled up in transit and 'burst open the vessel'. The seeds were rescued and put on another boat, which promptly capsized and sank. Undeterred, Maries set about gathering replacements. A further excursion led to his falling victim to extreme sunstroke, which he survived to travel to China, where he 'was often threatened and robbed of his baggage'. He nevertheless managed to send back some 500 living plants from his travels and introduced a number of trees and herbaceous plants, including firs (*Abies veitchii* and *A. mariesii*), maples and hydrangeas, lilies, primulas and spiraeas, and the Japanese squirrel's foot fern, *Davallia mariesii*. His herbarium was sent to Kew, and Sir Joseph Hooker (who had succeded his father, William, as director there) was sufficiently impressed by Maries's abilities that he recommended him to the Maharajah of Darbhanga in Bihar, where he laid out the palace grounds, going on to do the same for the Maharajah of Gwalior. He continued to collect plants in Darjeeling, sending seeds back to Kew, and became an expert on the cultivation of mangoes.

Père David was merely one of a large number of French missionaries in China who combined plant hunting with proselytism. When not preaching the gospel, these remarkable men discovered and described many plants, several of which bear their names. The *Incarvillea* genus is named for the eighteenth-century Jesuit Pierre d'Incarville (1706–57), sent to China in 1740 with the daunting task of attempting to convert the emperor himself to Christianity. Unsurprisingly, the emperor showed little interest in either the Bible or the missionary – until he saw a specimen of *Mimosa pudica* that d'Incarville had raised. The species had been given this name by Linnaeus because its leaves retract when touched, suggesting to the man who achieved notoriety for writing about the sexual parts of plants a natural modesty, *pudicus* being the Latin for chaste. Because of its apparent recoiling from human touch, this mimosa was also known as *noli me tangere* ('touch me not'), which is what the risen Christ said to Mary Magdalene on the first Easter morning, when, having taken him for a gardener, she realised who he was. It may have been for this religious reference that d'Incarville sent the plant to the emperor, perhaps thinking its name would prompt a discussion of Christ's resurrection; or he may have realised, quite correctly, that the behaviour of the plant would intrigue his potential convert. Whatever the case, the ruse worked for botany, if not for Christianity, because the emperor subsequently allowed the missionary into the Imperial Gardens and d'Incarville sent seeds of numerous Chinese native species to the Jardin des Plantes in Paris. He also explored the region and was responsible for the introduction of *Ailanthus altissima*, the tree of heaven, so named because it grows rapidly to a great height, although its pungent smell suggests the lower rather than the celestial regions; the Chinese pagoda tree, confusingly named by Linnaeus *Sophora japonica*; and the golden rain tree, *Koelreuteria paniculata*, so called because of its panicles of yellow flowers, and also known confoundingly as the Pride of India, a country in which it is not native.

The trumpet flowers of incarvilleas come in various colours, but the most popular species is the pink-purple *I. delavayi*, its name honouring one of d'Incarville's nineteenth-century successors, Père Jean Marie Delavay (1834–95). Reginald Farrer declared the plant a 'splendid find [...] which, in itself, is almost enough to reconcile oneself to the existence of missionaries'. Delavay had joined the Missions Étrangères de Paris at the south-eastern city of Huizho in 1867, and it was while he was on home leave in 1881 that he met Père David, who encouraged him to hunt specimens on his return to China and send them to the Natural History Museum in Paris. Weathering the plague, and undaunted by paralysis of the left arm, Delavay collected and sent back an astonishing 200,000 herbarium specimens, amongst which were more than 1,500 new species. Many of these were sorted and catalogued by Adrien Franchet (1843–1900), a leading botanist and taxonomist at the Natural History Museum, who published a 240-page *Plantae Delavayanae* (1889) and named several species after the indefatigable priest: the Chinese evergreen magnolia, *Magnolia delavayi*; the fragrant spring-flowering shrub *Osmanthus delavayi*; the densely foliaged privet *Ligustrum delavayi*; the Yunnan tree peony, *Paeonia delavayi*; and the delicate lilac Chinese meadow rue, *Thalictrum delavayi*, of which the 'Hewitt's Double' cultivar has become a garden favourite. Franchet himself is commemorated by the red-berried *Cotoneaster franchetii* and the tomato-red Chinese lantern, *Physalis alkek-engi* var. *franchetii*.

Luckily for gardeners, several other French missionaries found time away from their religious and relief work to travel and gather plants. The *Fargesia* bamboo genus is named to honour Paul Guillaume Farges (1844–1912), who worked in a mountainous region of north-east Sichuan from 1867 and began plant hunting in the early 1890s. He collected some 4,000 herbarium specimens and sent seeds to the leading French nursery Vilmorin, which had been established in Paris in 1743 and, like Veitch & Sons in England, was responsible for the introduction of innumerable plants in the west. In particular, the third-generation nurseryman Maurice de Vilmorin (1849–1918) saw the opportunities offered by botanically minded missionaries. He established an arboretum (which still exists) in the southern suburb of Verrières-le-Buisson, where he grew trees and shrubs from all around the world, many of them from seeds gathered by missionaries in China. It was Farges who sent a consignment of unviable-looking seeds of *Davidia involucrata* to Vilmorin, who nevertheless sowed them and managed to nurture a single seedling to maturity, thus introducing the plant to France at around the same time E. H. Wilson introduced it to Britain. Because Père David had been the first person to describe the handkerchief tree, Farges did not give his own name to the genus, but a large number of other plants bear his specific epithet, including the equally beautiful *Paulownia fargesii*. Among other plants named for Farges are the lilac-flowered *Catalpa fargesii*; the Chinese hazel *Corylus fargesii*, prized for its coppery bark; *Decaisnea fargesii*, a curiosity largely grown for its large, fat, slate-blue seedpods; Farges's cobra lily, *Arisaema fargesii*; and a decorative shrub with an English name, Farges's harlequin glorybower, that is almost as much of a mouthful as its Latin one, *Clerodendron trichotomum* var. *fargesii*.

Jean André Soulié (1858–1905) was a medical missionary working in China along the Tibet border, who disguised himself as a local, adopting Chinese dress and growing a long grey beard. This did not save him when the Tibetan revolt against the 1904 British invasion of Lhasa broke out in 1905. The lamas killed numerous missionaries and their converts: Soulié's chapel was burned to the ground, 200 converts were murdered, and the priest himself tortured and then shot. Like Farges, Soulié collected and sent large quantities of seeds back to Vilmorin, including the first ones of *Buddleja davidii*. In all he collected more than 7,000 species, and is commemorated by a large number of plants, among them the large, grey-leafed shrub rose *Rosa soulieana*, which has masses of small, single, clove-scented white flowers opening out of yellow buds. The specific epithet *souliei* also commemorates him in a variety of plants, including a catmint, *Nepeta souliei*, a white-flowered rhododendron, and the deep garnet-red, almost black, fritillaria-like *Lilium souliei*.

Alongside these French missionaries, several British plant hunters were, like Wilson, busy collecting plants in the Far East during the late nineteenth and early twentieth centuries. In 1905 the Scottish collector George Forrest, who has already been mentioned, had a very narrow escape from the rampaging Tibetans who had murdered Soulié, but survived an attack and a long journey to safety on foot and without boots, to discover in Yunnan a large number of shrubs and species of primula. Of the latter, he gave his own name to one species, *Primula forrestii*, while two more were named after his employer A. K. Bulley of Bees Nursery, *P. bulleyana* and *P. beesiana*. He subsequently fell out with Bulley and was snapped up by the politician and gardener J. C. Williams of Caerhays Castle in Cornwall, who had also sponsored E. H. Wilson and had a species of Chinese rhododendron, *R. williamsianum*, named after him. Of the many rhododendron species Forrest discovered, a red-flowered one has the name *Rhododendron forrestii*; this epithet is used too for many other shrubs and trees, including an acer, a pieris, a hypericum and a sorbus, as well as species of iris, hedychium (ginger) and pleione.

Equally intrepid, though having a very different temperament, was Reginald Farrer (1880–1920). Born with a hare-lip and a cleft palate, the latter inexpertly mended, Farrer was additionally very short, but did not allow these disabilities to deter him from becoming one of the foremost plant hunters in China and Tibet. He was witty, clever, arrogant, a terrible show-off, and could be astonishingly rude, but he inspired genuine, if often tested, affection in those who knew him well. Perhaps in compensation for an appearance that even his friends conceded was grotesque, and an emotional life desolated by unfulfilled homosexual yearnings, he cultivated an exquisite writing style that was authoritative, allusive, inventively descriptive, if occasionally arch, examples of which have already been given. As well as accounts of his travels, he published several books on rock gardening, which became his particular passion, notably his wonderful two-volume *The English Rock Garden* (1919). Like Forrest, Farrer would die in the field, succumbing to what may have been diphtheria in Burma, but several species of plants commemorate him with the epithet *farreri*, including a cypripedium and an alpine gentian discovered in China in 1914 and 1915 respectively, a lovely little purple allium, a pink alpine geranium, and the well-known, pale-pink and sweetly scented *Viburnum farreri*. Farrer (rightly) had a high opinion of his discoveries, and indeed himself, and he unblushingly described his own gentian as 'by far the most astoundingly beautiful of its race'. It inspired Farrer to one of his great flights of description: 'a single huge upturned trumpet wide-mouthed, and of an indescribably fierce luminous Cambridge blue within (with a clear white throat), while, without, long vandykes of periwinkle-purple alternate with swelling panels of nankeen, outlined in violet, and with a violet meridian line'. And if that doesn't inspire you to rush to the nearest nursery then nothing will.

The other great plant collector in the region, as well as in north-east India and Burma, was Frank Kingdon-Ward (1885–1958), commemorated by the species name *wardii*. Kingdon-Ward had abandoned a career as a schoolmaster in Shanghai to replace Forrest as A. K. Bulley's plant hunter, and he set off for south-west China in 1911. Another trip to the eastern Himalayas followed in 1913–14, and although his plant hunting was interrupted by war service in the Indian Army, he collected 97 different species of rhododendron and the first viable seeds of the blue Himalayan poppy, *Meconopsis betonicifolia*, during a trip to Tibet in 1924–5. The rhododendron that bears his name, *Rhododendron wardii*, has fragrant, pale yellow flowers, and among other plants with this epithet (not all of them from countries he explored) are a cotoneaster, several orchids, and *Anthopterus wardii*, a curious relative of the blueberry with scarlet flowers followed by edible berries of the brightest purple. He named *Lilium wardii*, which he found in Tibet and which has pink, carmine-stippled turk's-cap flowers, after himself, and the little pale-pink *Lilium mackliniae*, which he found in Manipur, north-east India, after his second and much younger wife, Jean Macklin, who (unlike her predecessor – see 'Atlas') accompanied him on several of his plant-hunting expeditions.

*

The western states of America did not get properly explored by plant hunters until 1804 when two American soldiers, William Clark and Meriwether Lewis, were sent by the government on an eighteen-month mission to discover and survey a transcontinental route from St Louis, Missouri, to the Pacific. Neither man was a botanist, but they were additionally tasked with observing and recording the natural history of the regions they explored. They saw plenty of plants and collected seeds on their return journey, many of them unknown to science. Among their discoveries was *Lewisia rediviva*, the taproots of which were sometimes dried and eaten by Native Americans, who called them bitter root – for reasons Lewis discovered to his disgust when he sampled them. He also named several local features after the plant, including the Bitterroot Mountains and the Bitterroot River, and lewisias are now grown for their rosette flowers which come in a wide range of colours. A purple monkey flower, *Mimulus lewisii*, and the Western mock orange, *Philadelphus lewisii*, both discovered on the expedition, are two more plants named for him, while his companion is commemorated by *Clarkia*, a genus of brightly coloured annuals they found growing by a river which now also bears Clark's name. One of those who took a keen interest in the expedition was Bernard M'Mahon (1775–1816), who had emigrated to America from his native Ireland and established a nursery just outside Philadelphia. He published America's first seed catalogue, and his *The American Gardener's Calendar* (1807) was the first practical book on American gardening. He introduced one of the shrubs brought back from the Lewis and Clark expedition into cultivation and it now bears his name: *Mahonia*.

Later plant hunters in both North and Central America include two German brothers, Carl (1851–1941) and Josef (1860–1932) Purpus. Gardeners might be puzzled by the species name of *Lonicera × purpusii*, a winter-flowering shrub with very fragrant cream-coloured flowers. The name is close to similar-sounding epithets meaning purple (see 'Spectrum'), but the plant was in fact one of several named after the brothers, including the Snow Mountain penstemon, *Penstemon purpusii*, an alpine species native to northern California.

Naming plants after those who go searching for them continues to this day, with such epithets as *lancasteri* (for Roy Lancaster, b. 1937) and *grey-wilsonii* (for Christopher Grey-Wilson, b. 1944) entering the botanical dictionaries. Others who have plants named after them include, as we have seen, owners of nurseries and gardens. Nurserymen such as Arthur Burkwood (1888–1951) of the Parkwood Nurseries in Kingston-upon-Thames and Walter Ingwersen (1883–1960) of Birch Farm Nursery in Surrey are remembered in the names of such plants as *Viburnum × burkwoodii* and *Sempervivum ingwersenii*, while Sir Harold Hillier (1905–85) and his family's horticultural business lend their names to such hybrids as *Eucryphia × hillieri* and × *Halimiocistus wintonensis*, the latter epithet derived from Wenta, the ancient name for Winchester, where the business started. One of Hillier's employees, Eric Smith (1917–86), spent much of his time hybridising bergenias, hostas and hellebores, and *Helleborus × ericsmithii* was one of the results – the epithet chosen because several earlier Smiths had already nabbed *smithii*, *smithianus*, and *smithiae*. Nowadays nursery owners tend to give their names – or those of their wives and daughters (sons, for some reason, rarely getting a look-in) – to cultivars, but these are not Latinised. Roses and sweet peas in particular are offered for sale under the names of well-known people, from Princess Diana to Alan Titchmarsh, but these are (thankfully) beyond the scope of this book.

Although the names of such famous and influential gardeners as William Robinson (1838–1935), Gertrude Jekyll (1843–1932) and Vita Sackville-West (1892–1962) have also been given to cultivars, you will not find them commemorated in any species names. There are, however, species named after other and less well known gardeners such as Edward Whittall (1851–1917), who was born into a family of wealthy merchants in what was then Smyrna (now Izmir) in Turkey, where he created a garden of local plants which still exists and can be hired for events. Whittall had become interested in growing native species after arranging a shooting trip for botanist, lepidopterist and hunter of game and plants Henry John Elwes (1846–1922), who collected lilies in the Himalayas and Korea, and would go on to write both a monograph on the genus and, with Augustine Henry, a seven-volume *Trees of Great Britain and Ireland* (1906–13). Enthused by Elwes and his interest in Turkish flora (including the giant snowdrop he collected there and named *Galanthus elwesii*), Whittall collected bulbs from the region, huge numbers of which he sent to Kew. Among the species by which he is commemorated are *Fritillaria whittallii*, the flowers of which are green with purple chequerboard markings, the small dark orange *Tulipa whittallii*, the pale lilac-blue *Gentiana whittallii* and a *whittallii* variety of Elwes's snowdrop. His beneficiaries at Kew have their own species name, mostly used for hybrids such as the pale, creamy yellow Kew broom, *Cytisus* × *kewensis*, the lemon-yellow Alpine primrose *Primula* × *kewensis*, and *Magnolia* × *kewensis*, the 'Wada's Memory' cultivar of which is in early spring so smothered in white, elegantly lax flowers that you can barely see the actual tree at all.

Several generations of the celebrated Rothschild family have been keen horticulturists, and the family's banking fortune, founded in Frankfurt in the eighteenth century, has allowed the creation of a number of large and important gardens, both in Britain and in Europe. Among Rothschild plants is the extremely rare orchid *Paphiopedilum rothschildiana*, 'the most spectacular species in the genus', named in honour of Ferdinand de Rothschild (1839–1898), who built Waddesdon Manor in Buckinghamshire. The majority of plants bearing the family epithets, however, are named for Walter Rothschild (1868–1937), who had studied zoology at university and created his own natural history museum at Tring Park in Hertfordshire, where he also cultivated orchids and employed people to hunt them in remote regions. Among exotic plants that bear his name are *Vanda rothschildiana*, a hybrid orchid with strongly marked violet flowers, and the red-and-yellow climbing flame lily, *Gloriosa rothschildiana*. Walter's younger brother Charles Rothschild (1877–1923) lived at Aston Wold in Northamptonshire, where *Rosa* × *rothschildii* was discovered. Although he worked in the family bank, Charles was an expert entomologist and a pioneering conservationist, who would run Ashton Wold more or less as a nature reserve, a lead followed by his daughter Miriam (1908–2005), who became a famous champion of British wild flowers as well as an expert on fleas. Charles undertook entomological and botanical expeditions, and the natural hybrid *Iris* × *rothschildii*, discovered in Hungary, was also named after him. His cousin Lionel Rothschild (1882–1942) created Exbury Gardens in Hampshire, where he grew huge numbers of rhododendrons, azaleas and camellias, and so it seems right that a species of rhododendron discovered in Yunnan in 1972 should be named *Rhododendron rothschildii*.

Although not as rich as the Rothschilds, the eccentric Ellen Willmott (1858–1934), who created Warley Place in south-west Essex, was very well off indeed. At the age of seven she had received the first of several cheques for £1,000 from her indulgent godmother, apparently left on her breakfast plate, and when she was thirty-three she came into a fortune of £210,000. In her glory days she never wanted for money, all of which and more she spent on buying plants and properties, including a chateau in France and a large garden near Ventimiglia in Italy. She was described by Gertrude Jekyll in 1897 as 'the greatest living woman gardener', and she had some sixty species named after herself or after Warley Place. A leading figure in the RHS, she also sponsored a number of plant-hunting expeditions, including one to China by E. H. Wilson; among plants she grew from seed he brought back from his travels are the blue, late-flowering *Ceratostigma willmottianum*, the magenta single-flowered *Rosa willmottiae* and the Chinese witch hazel *Corylopsis willmottiae*. Plants named after Warley Place, where she employed 100 gardeners, include *Epimedium* × *warleyense* and the Central Asian *Iris warleyensis*. 'My plants and my gardens come before anything in life for me,' she admitted, and that included financial prudence. Her extravagance eventually outran her fortune, and in order to pay off her creditors she was obliged to divest herself of most of her possessions, including her two foreign properties. After she died, her house was sold to settle her outstanding debts and then demolished. Nature gradually reclaimed the garden, which is now maintained as a nature reserve by the Essex Wildlife Trust, although some of the thousands of bulbs she planted still come up every year, notably the daffodils she so loved, and for which she won several RHS gold medals. In many ways this is a sad story, like many of those that have gone before, and yet Ellen Willmott lives on in the species named after her. She would no doubt have been pleased, but possibly even more so that she is best remembered not for a species but for a plant nickname, 'Miss

Wilmott's ghost', given to the tall, silver sea holly *Eryngium giganteum*. She carried seeds of the plant in her pockets and surreptitiously scattered them in other people's gardens when she visited them, a cheeky bid for immortality that no taxonomist could later thwart.

Word Lists

GENERA

Achillea – Achilles, Greek warrior hero in the *Iliad*

Adonis – Adonis, mortal lover of Aphrodite (Venus)

Asclepia – Asclepios, Greek god of medicine

Atropa – Atropos, one of the three Fates of Greek mythology

Aubrieta – Claude Aubriet (1665–1742), French botanical illustrator at the Jardin des Plantes

Banksia – Sir Joseph Banks (1743–1820), plant collector and supervisor of Kew Gardens

Bauera – Ferdinand (1760–1826) and Franz (1758–1840) Bauer, Austrian botanical illustrators

Bauhinia – Johann (1541–1613) and Gaspard (1560–1624) Bauhin, Swiss botanists

Begonia – Michel Bégon (1638–1710), French colonial administrator and patron of botanical expeditions

Bignonia (and specific *bignonioides*) – Jean-Paul Bignon (1662–1743), librarian to Louis XIV

Bocconia – Paolo Boccone (1633–1704), Italian monk and physician in Sicily

Bougainvillea – Louis-Antoine de Bougainville (1729–1811), French admiral and explorer in the Pacific

Brunnera – Samuel Brunner (1790–1844), Swiss military doctor and naturalist

Buddleja – Rev. Adam Buddle (1662–1715), clergyman and botanist

Camellia – Georg Josef Kamel (1661–1706), Moravian Jesuit pharmacist working in the Philippines

Carpenteria (and specific *carpenteri*) – William Carpenter
(1811–48), Louisiana doctor, botanist and geologist

Centaurea – Chiron, the centaur who taught Achilles and Ajax

Clarkia – William Clark (1770–1838), American soldier and
plant hunter

Commelina – Caspar Commelijn (1668–1731), Dutch botanist
and author of a treatise on earthworms

Cypripedium – Aphrodite, Greek goddess also known as Cypris

Dahlia – Anders Dahl (1751–89), Swedish doctor and botanist,
author of a 1787 commentary on Linnaeus's system of
classification

Daphne – Daphne, water nymph

Deutzia – Johan van der Deutz (1743–88), Amsterdam lawyer
and sponsor of Thunberg

Eschscholzia – Johann von Eschscholtz (1793–1831), Baltic
German doctor and naturalist

Eupatorium – Mithridates Eupator, King of Pontus (135–63 B C)

Euphorbia – Euphorbus, late-first-century B C Greek physician

Forsythia – William Forsyth (1737–1804), horticulturist and
founder member of the RHS

Fothergilla – John Fothergill (1712–80), Quaker doctor and founder
of a botanical garden at Upton House in West Ham, London

Fremontodendron – John C. Frémont (1813–90), US explorer,
soldier and politician

Fuchsia – Leonhart Fuchs (1501–66), German doctor and
botanist, founder of the Tübingen Botanical Garden (1535)
and author of the illustrated herbal *De historia stirpum
commentarii insignes* (1542)

Gardenia – Alexander Garden (1730–91), Scottish-born
American doctor and naturalist

Garrya – Nicholas Garry (c. 1781–1856), trader with the
Hudson Bay Company

Gentiana – Gentius, King of Illyria (fl. 181–168 B C)

Gunnera – Johan Gunnerus (1718–73), Norwegian botanist
    and bishop

Hacquetia – Balthasar Hacquet (1739/40–1815), Austrian physicist

Helenium – Helen of Troy

Helianthus, Helianthemum – Helios, Titan and personification
    of the sun in Greek mythology

Heuchera – Johann Heinrich von Heucher (1677–1746),
    German professor of botany and medicine, author of
    the first published catalogue of plants growing in the
    Wittenberg Botanical Garden

Hosta – Nicolaus Tomas Host (1761–1834), Austrian botanist
    and doctor, first director of the botanical garden at the
    Belvedere Palace

Houttuynia – Maarten Houttuyn (1720–98), Dutch naturalist

Hyacinthus – Hyacinthus, beautiful youth loved by Apollo

Iris – Iris, messenger of the gods in Greek mythology

Kerria – William Kerr (d. 1814), Scottish gardener and plant
    hunter

Knautia – Christof (1638–94) and Christian (1654–1716)
    Knaut, German botanists and doctors

Kniphofia – Johann Hieronymus Kniphof (1704–63), German
    professor of medicine, whose 10-volume *Botanica in
    originali seu herbarium vivum* (1759–64) was the first
    botanical atlas to use the Linnaean system

Koelreuteria – Joseph Kölreuter (1733–1806), German scientist
    and pioneer in hybridisation and the role of insects in
    pollination

Lapageria – Joséphine Bonaparte (1763–1814), *née* de la
    Pagerie, wife of Napoleon

Lavatera – Jean Rodolphe Lavater (fl. 1700–16), Swiss doctor
    and botanist

Lewisia – Captain Meriwether Lewis (1774–1809), soldier
and explorer in the American West with William Clark
[see above]

Libertia – Marie-Anne Libert (1782–1865), Belgian botanist
and mycologist

Liriope – water nymph, mother of Narcissus

Lobelia – Matthias de l'Obel (Lobelius) (1538–1616),
Flemish physician and botanist who settled in England in
1596, becoming personal physician and royal botanist to
James I.

Lonicera – Adam Lonitzer (Lonicerus) (1528–86), German
author of *Naturalis historiae opus novum* (2 vols, 1551, 1555)
and its German translation, the *Kräuterbuch* (1557)

Lysimachia – Lysimachus (c. 360–281 B C), ruler of Thrace

Macleaya – Alexander Macleay (1767–1848), Scottish
lepidopterist and botanist

Magnolia – Pierre Magnol (1638–1715), director of the
Montpellier Botanic Garden

Mahonia – Bernard M'Mahon (1775–1816), Irish-born
horticulturist, nurseryman and author in Philadelphia

Matteuccia – Carlo Matteucci (1811–68), Italian physicist

Matthiola – Pietro Andrea Mattioli (Matthiolus) (1501–77),
Sienese botanist and author who also first described cat
allergy

Monarda – Nicolás Monardes (1493–1588), Seville-born Italian
doctor to Philip II of Spain and author of first book about
plants from the Americas

Morina – René and Pierre Morin, Parisian nurserymen
from 1617, specialising in bulbs, who corresponded and
exchanged plants with the Tradescants

Musa – Antonius Musa (late first century B C), doctor to
Emperor Augustus

Narcissus – Narcissus, Greek youth who fell in love with his own reflection

Nicotiana – Jean Nicot (1530–1600), French ambassador to Portugal

Paeonia – Paeon, physician to the Greek gods

Paulownia – Anna Pavlovna (1795–1865), daughter of Tsar Paul I of Russia who narrowly missed being married off to Napoleon, instead becoming the wife of William II of the Netherlands

Perovskia – Count Vasily Perovsky (1794–1857), Russian general and statesman

Philadelphus – (possibly) Ptolemy II Philadelphus of Egypt (308–246 B C)

Pleione – Pleione, mother of the Pleiades

Puschkinia – Count Apollo Mussin-Puschkin (1760–1805), Russian mining expert and plant collector in the Caucasus

Robinia – Jean Robin (1550–1629), royal gardener to the French kings Henri IV and Louis XIII

Rodgersia – John Rodgers (1812–82), American naval officer, who found *R. podophylla* in Japan

Romneya – Thomas Romney Robinson (1792–1882), Irish scientist and astronomer at the Armagh Observatory

Scopolia – Antonio Scopoli (1723–88), Italian physician, botanist and geologist

Smyrnium – Myrrha, incestuous mother of Adonis

Stewartia – John Stuart, 3rd Earl of Bute (1713–92), British prime minister, keen botanist and patron of horticulture (Named by Linnaeus, who got the spelling wrong.)

Stokesia – Jonathan Stokes (1755–1831), English botanist

Strelitzia – George III's wife, Charlotte of Mecklenburg-Strelitz (1744–1818)

Sutherlandia – James Sutherland (1639–1719), Scottish curator

of both Trinity Hospital Garden and Edinburgh University
Physic Garden, which he catalogued (1683)

Tolmiea (and specific *tolmiei*) – William Tolmie (1812–86),
Scottish surgeon who worked for Hudson Bay Company
and later became a Canadian politician

Tulbaghia – Ryk Tulbagh (1699–1771), who joined the
Dutch East India Company at sixteen and ended up
Governor of Dutch Cape Colony. Corresponded with
Linnaeus and sent him 200 specimens of South African
plants

Tweedia – James Tweedie (1775–1862), head gardener at Royal
Botanic Garden Edinburgh, who emigrated to Buenos
Aires in 1825 and travelled in South America, sending back
seeds to William Hooker

Waldsteinia – Franz von Waldstein (1759–1823), Austrian
soldier, explorer and botanist, whose younger brother was a
patron of Beethoven

Weigela – Christian Weigel (1748–1831), German scientist,
chemist and botanist

Wisteria – Caspar Wistar (1761–1818), professor of anatomy at
the University of Pennsylvania

Zantedeschia – Giovanni Zantedeschi (1773–1846), Italian
physician and botanist

Zinnia – Johann Gottfried Zinn (1727–59), Bavarian botanist,
physician and anatomist

SPECIES

battandieri – Jules Battandier (1848–1922), French botanist
who wrote several books on the flora of Algeria

beesianus – Bees nursery (founded 1903)

bodnantensis – Bodnant Garden, Wales (founded 1874)

borisii – Boris III of Bulgaria (1894–1943), monarch and plant
 hunter
bowlesianus – E. A. Bowles (1865–1954), English plantsman
 and writer
bulleyanus – A. K. Bulley (1861–1942), Liverpool cotton
 merchant, patron of plant hunters and founder of both Ness
 Botanic Gardens and Bees Nursery
bungei, bungeanus – Alexander Bunge (1803–90), Russian-
 German botanist and plant collector
burkwoodii – Arthur (1888–1951) and Albert (1890–?)
 Burkwood, British nurserymen
catesbaei – Mark Catesby (1683–1749), British naturalist,
 explorer, author and painter
clusiana, clusii – Charles de l'Écluse (Clusius) (1526–1609),
 French doctor, botanist and writer born in Leiden
coulteri – Thomas Coulter (1793–1843), Irish botanist and
 curator of Trinity College Dublin herbarium
darwinii – Charles Darwin (1809–82), author of *On the Origin
 of Species*
dombeyi – Joseph Dombey (1742–94), French botanist and
 explorer
drumondii – Thomas Drummond (1793–1835), Scottish plant
 collector in America
elwesii – Henry John Elwes (1846–1922), British plant hunter in
 the Himalayas and Korea.
ericsmithii – Eric Smith (1917–86), English nurseryman
farreri – Reginald Farrer (1880–1920), English plant hunter
 and author
fortunei – Robert Fortune (1812–80), Scottish plant collector in
 Far East
franchetii – Adrien Franchet (1843–1900), leading botanist and
 taxonomist at the Natural History Museum in Paris

frikartii – Karl Frikart (1879–1964), Swiss plantsman and hybridiser

grey-wilsonii – Christopher Grey-Wilson (b. 1944), British alpine plant specialist

haastii – Johan Julius von Haast (1824–87), German geologist and museum director in New Zealand

hartwegii – Karl Theodor Hartweg (1812–71), German botanist sent by the RHS to collect plants in Mexico, South America and California

heldreichii – Theodor von Heldreich (1822–1902), German botanist, plant collector and author, director of the Royal Garden in Athens

hemsleyanus – William Hemsley (1843–1924), English botanist who trained at Kew and catalogued the flora of China

hendersonii – Louis Forniquet Henderson (1853–1942), American botanist and plant collector

henryi – Augustine Henry (1857–1930), British customs officer and plant collector in China and Taiwan

hillieri – Hillier's Nursery, Hampshire, founded 1864

hookeri, hookerianus – William Jackson Hooker (1785–1865), explorer, plant collector, author and director of the Royal Botanic Gardens, Kew, and his son and successor at Kew, Joseph Dalton Hooker (1817–1911)

ingwersenii – Walter Ingwersen (1883–1960), English nurseryman

jacquemontii, jacquemontianus – Victor Jacquemont (1801–32), French naturalist who explored India, the Himalayas, Kashmir and Tibet

jeffreyi – John Jeffrey (1826–54), Scottish botanist and plant hunter working in the USA, who disappeared while crossing the Colorado Desert

kaempferi – Engelbert Kaempfer (1651–1716), German doctor
working for Dutch East India Company in Japan, author of
a *History of Japan* (1727)

kewensis – Royal Botanic Gardens, Kew

lamarckii – Jean-Baptiste Lamarck (1744–1829), French
zoologist and botanist who coined the word vertebrate

lancasteri – Roy Lancaster (b. 1937), English plant hunter,
broadcaster and author

langsdorfii – Georg Langsdorf (1774–1852), German physician
and botanist

lawsoniana – Charles Lawson (1795–1873), Edinburgh
nurseryman specialising in grasses and conifers, including
the Lawson cypress *Chamaecyparis lawsoniana*

leichtlinii – Max Leichtlin (1831–1910), German horticulturist
with a garden in Baden-Baden

leylandii – C. J. Leyland (1849–1926), landowner who bred
conifers at Haggerston Hall, Northumberland, including ×
Cupressocyparis leylandii

lindheimeri – Ferdinand Lindheimer (1801–79), German-Texan
botanist

lyallii – David Lyall (1817–95), Scottish botanist and
explorer who travelled with James Ross to Antarctica in
1839, bringing back some 1,500 herbarium specimens.
Subsequently collected in New Zealand and Canada

maackii – Richard Maack (1825–86), Russian naturalist
working in Siberia and eastern Asia

magellanicus – Ferdinand Magellan (1480–1521), Portuguese
mariner; or native to Southern tip of South America

makinoi – Tomitaro Makino (1862–1957), known as the 'Father
of Japanese Botany'

maximowiczii – Karl Maximowicz (1827–91), Russian
botanist and director of the St Petersburg Botanical

Gardens, who travelled in Far East and described over
2,300 new plants

meadius – Richard Mead (1673–1754), English doctor and royal
physician to George II

mlokosewitschii – Ludwik Młokosiewicz (1831–1909), Polish
army officer and plant collector

moyesii – James Moyes (1876–1930), British clergyman with
the China Inland Mission

northiae – Marianne North (1830–90), British botanical artist
and traveller, whose collection of paintings is displayed in
its own gallery at Kew

nuttallii – Thomas Nuttall (1786–1859), English plant collector
in America who brought plants back for the Liverpool
Botanical Garden. Author of *The Genera of North American
Plants* (1818)

pennaei – Thomas Penny (c. 1530–89), English botanist

pernyi – Paul-Hubert Perny (1818–1907), French missionary in
Guizhou Province who supplied plants to Adrien Franchet
and introduced the tussar silk moth (*Bombyx pernyi*) to France

przewalskii – Nikolai Przewalski (1839–88), Russian
geographer working in Central Asia

purpusii – Carl (1851–1941) and Josef (1860–1932) Purpus,
German plant hunters in America

rehderianus – Alfred Rehder (1863–1949), German-American
dendrologist at the Arnold Arboretum

renardii – Karl Renard (1809–86), German scientist of
French origin working in Russia, director of Moscow's
Zoological Museum and president of the Imperial Society
of Naturalists

requienii – Esprit Requien (1788–1851), French naturalist who
founded Avignon's Natural History Museum, where he
created a herbarium of 300,000 specimens

rothschildianus, rothschildii – Walter Rothschild (1868–1937),
  naturalist, and other members of the Rothschild family

schubertii – Gotthilf von Schubert (1780–1860), German
  philosopher and naturalist

seemannia – Berthold Seemann (1825–71), Kew-trained
  German plant collector in the Pacific and South America

sieberi, sieberianus – Franz Sieber (1789–1844), Prague-born
  botanist and plant collector who spent the last fourteen
  years of his life in a lunatic asylum

sieboldii, sieboldiana – Philipp Franz Balthasar von Siebold
  (1796–1866), German botanist, doctor and explorer
  working in Japan

sprengeri – Karl Ludwig Sprenger (1846–1917), deaf German
  botanist who established a nursery in Naples, decimated by
  volcanic ash after eruption of Vesuvius in 1906

stellerianus – Georg Wilhelm Steller (1709–46), German
  botanist and zoologist in Russia and Alaska

stewartii – Laurence Stewart (1877–1934), Keeper of the Royal
  Botanic Garden Edinburgh

thomsonii – Thomas Thomson (1817–78), Scottish doctor and
  botanist working in India

tommasinianus – Muzio Tommasini (1794–1879), Trieste-born
  botanist and politician who went on botanical trips in
  Europe

tournefortii – Joseph Pitton de Tournefort (1656–1708), French
  botanist, traveller and writer

turczaninowii, turczaninovii – Nicolai Turczaninov (1796–
  1863), Russian botanist

wallachianus, wallachii – Nathaniel Wallich (1786–1854),
  Danish botanist and doctor working in India.

wardii – Frank Kingdon-Ward (1885–1958), British plant
  collector

watereri – Waterers Nursery, Bagshot, Surrey since 1829

whittallii – Edward Whittall (1851–1917), merchant, plant
collector and garden owner in Turkey

williamsianus – J. C. Williams (1861–1939), British politician
and gardener of Caerhays Castle, Cornwall

wilmottianus – Ellen Willmott (1858–1934), British gardener of
Warley Place, Essex

wintonensis – Hillier's Nursery, Hampshire, founded 1864

MISCELLANEA

Seasons, shapes and sizes

When advertising the plants they hope we will buy, nurseries have a fairly standard descriptive list, the essentials of which will be printed in a catalogue or crammed on to labels. There will be a botanical name; information as to whether the plant is annual, biennial or perennial; the season in which it will flower (if it does), and the colour and shape of those flowers; if relevant, details of the foliage and any change of colour it goes through; the growing conditions in which it will thrive; its eventual height and spread. Most of these attributes are among those that also help botanists determine which genus or species a plant belongs to. Clearly a species name that describes a plant's habit – whether it is tall or short, whether it thrusts upright or spreads at ground level – or the season in which it flowers or is otherwise at its most characteristic or best, can be helpful when planning a border, but this does not apply to all the names that refer to details of leaves, flowers, fruits and other parts. The exact shape or texture of an individual leaf, for example, what kind of stalk attaches it to a stem and how it is arranged there, or minute differences in the structure of the flower, are really of more interest to scientists trying to sort out one species from another than to the rest of us who are merely dithering over which plant to buy at the local garden centre – fascinating though some of these fine details may be. Whereas botanists may look at *Buddleja alternifolia* and decide that what distinguishes it from other buddleias is that its leaves 'grow from alternating points of a stem rather than opposite each other', for the rest of us it is distinguished by the fact that its purple flowers are borne all along those stems, which have a gracefully weeping habit. That said, although I originally planted *Allium triquetrum* in my garden because it flowers early in the spring, and those white flowers have prominent mustard-coloured anthers and a handsome dark-green stripe down the centre of every petal, I am pleased to have learned that it gets its species name from the Latin *triquetrum*, meaning triangular, because of its unusual three-sided flower-stalks. Similarly, I grow *Geranium macrorrhizum*, in both the original magenta-pink form and a white cultivar, for its flowers and the pungent-smelling foliage that takes me straight back to Bulgaria, where it is a native and often grown in tubs outside shops and restaurants; but I am glad to know that its species name refers to the large, shining-brown root-stock that spreads visibly above ground. As for cyclamen, I find it interesting that the plant gained its name from the rounded shape of the tubers, rather than

Iberian cyclamen, *Cyclamen vernum*

the beautiful flowers and leaves for which we grow them: from *kyklos*, the Greek for circle.

Because early botanists often disregarded flowers when deciding upon names, many species within a genus are differentiated by epithets that describe the shape of their leaves. As will be seen by the word lists at the end of this chapter, leaf types far outweigh flower types. The RHS's *Index of Garden Plants* illustrates thirty-one basic leaf shapes, most of which have Latinate names such as ensiform (from *ensis* = sword), flabellate (from *flabellum* = fan) or cuneate (from *cuneus* = wedge). In addition there are thirty-two adjectives (such as acuminate and sagittate) to describe either the tips or the bases of leaves, and another twenty-three (such as crenate and fimbriate) describing leaf margins. These words are unlikely to be used on a daily basis by gardeners, and although many of them have their equivalents in species names, these are the kind of epithets that we can mostly relegate to word lists rather than discuss in any detail. Botanical names that distinguish plants by the colour of their flowers or foliage have already had a chapter to themselves, and those derived from the way a plant smells or feels to the touch were dealt with in 'Anatomy'. We will concentrate here on plants named because of some other distinguishing feature of their leaves, flowers or habit, or for the seasons in which they bloom. We'll also look at a number of more general specific epithets, such as *amabilis* (lovely), *blandus* (charming) and *elegans* (elegant), which seem too vague or subjective to be of much help to gardeners but have their own intrinsic interest.

The seasons in which plants flower are often embodied in botanical names, ranging from the rather vague *praecox*, meaning very early, to the highly specific *majalis*, meaning in the month of May. Naturally enough, plants do not always behave as their names promise, with weather, position, soil conditions and other factors delaying or precipitating flowering. Human intervention can also play a part; anyone who has been to the Chelsea Flower Show or the Malvern Spring Festival will know that plants can be brought early into flower for the purposes of display or to tempt buyers. It is nevertheless handy to know the progress of the seasons as they feature in species names. The Roman seasons were *ver* (spring), *aestas* (summer), *autumnus* (autumn) and *hiems* or *bruma* (winter), and these give us the specific epithets *vernus*, *vernalis* and *veris* for the spring, *aestivus* and *aestivalis* for the summer, *autumnalis* for the autumn, and *hiemalis* and *brumalis* for the winter. These mostly refer to when a plant flowers, but *autumnalis* sometimes denotes a plant that has interesting autumn foliage. Leucojums are a good example of plants that take both their English and botanical names from the different seasons: *Leucojum vernum* is the spring snowflake, *L. aestivum* the summer snowflake, and *L. autumnale* the autumn snowflake – or at any rate it used to be, but has now been shifted to another genus and is *Acis autumnalis*. The RHS also lists a winter snowflake native to France, *L. hiemale*, but this name is only 'tentatively accepted', which may mean that it has yet to be determined whether or not it is a separate species. This all sounds reasonably simple and satisfactory, but there is a climatic complication because, as Stearn notes, in southern Britain the species most gardeners grow, *L. aestivum*, tends to flower in the spring, between March and May; but it was named by Linnaeus, in whose native Sweden it does not flower until the summer, usually in June.

Working our way through the flowering year, we start with *praecox*, a Latin word meaning ripening before its time or maturing too soon and the root of the English word 'precocious'. *Stachyurus praecox* is a small tree that produces stiffly hanging racemes of pale yellow flowers along its boughs in February (its genus name is from the Greek, meaning 'spike-tailed'). Similarly the highly scented flowers of wintersweet, *Chimonanthus praecox*, appear well in advance of spring, sometimes as early as December, and indeed the genus name is from the Greek *cheimon*, meaning winter. The yellow broom *Cytisus* × *praecox* flowers considerably later in the spring, but it does so before other brooms. The word 'apricot' (originally 'apricock') is also related to *praecox*, as pedantically explained in an amusing scene in the 2017 film *Call Me by Your Name*. The apricot is one of the earliest flowering fruit trees, often in bloom in Britain in February, although its botanical name, *Prunus armeniaca*, refers to its presumed Armenian origins. The early spring flowering of the cowslip, *Primula veris*, is signalled in both genus and species name: *Primula* from the Latin *primus*, meaning first, and *veris* meaning of the spring – though it in fact generally flowers later than the English primrose, *P. vulgaris*. The genus name of these two plants is also used as a specific epithet, as in *Rosa primula*, the wonderfully scented incense rose, which produces its pale yellow flowers in the spring, way ahead of the first rose of summer. Spring-flowering plants that take other specific epithets include the spring crocus, *Crocus vernus*, which in its true form has what Reginald Farrer describes as 'a delicate purple bloom' but is more familiar in what E. A. Bowles calls 'fat, prosperous, gone into trade and done well with it, garden forms'. The spring pasqueflower, *Pulsatilla vernalis*, has delicate white flowers flushed blue-violet on the outside, whereas *P. vulgaris*, which as its name suggests was once a common British wild flower, has rather more robust purple ones. These plants' genus name indicates their family relationship to anemones, *Pulsatilla* being a diminutive of the Latin *pulsatus*, meaning beaten about, the notion being that these are little flowers that quiver in the wind. The English name suggests Easter flowering (from the Old French *pasque*), but Easter is of course a festival that moves around a good deal, the earliest date being 22 March, the latest 25 April. In any case, the name appears to be the invention of Gerard, before whose intervention the plant was called the passeflower, also derived from a French name, *passefleur*, the origins of which are obscure but may be connected with the plant's root being boiled in *passum*, a kind of Roman raisin-wine, and used (by Dioscorides and others) to cure eye troubles. By Gerard's time, the plant's 'vertues' had been discounted, and Gerard writes: 'They floure for the most part about Easter, which hath mooued mee to name it *Pasque Floure*.'

Next in the year comes the lily of the valley, *Convallaria majalis*, the genus name of which designates its natural habitat (from the Latin *valles* = valley) and its species name (from *maius* = May) its flowering time. The vigorous, deep pink, single-flowered *Rosa majalis* is another rose noted for its early flowering; although popular into the nineteenth century the cinnamon rose is now more often grown commercially because its hips are a particularly rich source of Vitamin C.

Defining plants as summer flowering seems somewhat invidious, given that this is the principal flowering season in the garden. The summer epithets are most often used to

Lily of the valley, *Convallaria majalis*

distinguish plants from other species in the genus that flower
at different times of the year, as with the leucojums. The spring
pheasant's eye, *Adonis vernalis*, for example, has bright yellow
flowers, but the more familiar species is the bright scarlet,
summer-flowering *A. aestivalis*.

The late-flowering equivalent of *praecox* is *serotinus*, a Latin
word meaning belated. We think of narcissi as spring flowers,
but the southern European *Narcissus serotinus*, which looks
rather like *N. poeticus* but with a tiny corona, flowers in the
autumn. *Iris serotina* similarly flowers later than other species,
and *Leucanthemella serotina* is an autumn-flowering species
of the ox-eye daisy. As summer fades, the autumn crocuses,
Colchicum autumnale, appear. They in fact belong to a different family from the spring crocus, and are also known as the
meadow saffron and (because the flowers appear before any
foliage) naked ladies or, as Gilbert White records in his 1791
journal, 'Naked Boys'. We think of heleniums as plants that
extend the flowering season, and most of those we grow in our
gardens are cultivars of *Helenium autumnale*.

Winter, being winter, has rather fewer plants with this
season's specific epithets. *Begonia* × *hiemalis* is a hybrid spe-
cially developed to flower during the winter months, and the
snow camellia, *Camellia hiemalis*, flowers in late autumn and
throughout the winter. Like the camellia, the *Chionodoxa*
genus of early-flowering bulbs has a popular name that refers
to a familiar feature of the season, glory-of-the-snow, but this
is simply a translation of its botanical name, from the Greek
chion = snow + *doxa* = glory. The epithet *brumalis* is even less
frequently used than *hiemalis*, though there is a species of
winter fungus, *Polyporus brumalis*, and a wattle, *Acacia brumalis*, which flowers during what are the winter months in its
native Australia. A further seasonal epithet is *hibernus*, another

Latin word for 'wintry' – the Roman army called its winter quarters *hiberna*, and the associated verb gives us 'hibernate'. *Austrocactus hibernus* has unusual caramel-coloured flowers, is native to South America, and gets its name because it is a frost-hardy species. This epithet should not be confused with *hibernicus*, which means from Ireland.

In 1749 Linnaeus made a study of the time of day flowers opened and closed, with the intention of creating a floral clock 'by which one could tell the time, even in cloudy weather, as accurately as by a watch'. He took particular note of species of hawksbeard (*Crepis*) and hawkbit (*Leontodon*), which may indeed have proved reliable timekeepers; however, some plants that open in the morning and close at noon or in the evening, such as the scarlet pimpernel (*Anagallis arvensis*) and the Venice mallow (*Hibiscus trionum*), only do so in sunny weather. Common names are often more suggestive than botanical ones in this matter. It is well known, for example, that the daisy

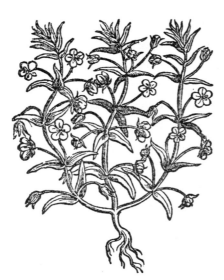

got its English name, a corruption of 'day's eye', because it opens in the morning and closes in the evening – and its medieval Latin name was *solis oculus*, the sun's eye. Other plants with English names that tell us when their flowers open and close include morning glory (*Ipomoea tricolor*), Jack-go-to-bed-at-noon (*Tragopogon pratensis*) and evening primrose (the

Scarlet pimpernel, *Anagallis arvensis*

Oenothera genus). The flowers of the evening primrose last less than twelve hours, usually fading by noon on the day after they bloom, and those of the day lily are similarly ephemeral, hence its common name. In this case the plant's botanical name, *Hemerocallis*, means something similar, derived from the Greek *hemera* = day + *kallos* = beauty.

Some dictionaries list the species name *matutinalis*, from the Latin *matutinus*, meaning 'of the early morning', but it seems not to be in current use. There are, however, the epithets *meridianus* and *meridionalis* for plants with flowers that open at noon, such as the Australian protea *Leucadendron meridianum* and the globe daisy *Globularia meridionalis*, the latter described by Farrer as 'useful if not brilliant little shrubby plants carrying ... balls of pale grey-blue fluff'. Hesperus is the Greek word for the evening star, and so both *Hesperantha* and *Hesperis* are plants associated with the latter part of the day. *Hesperantha* (syn. *Schizostylis*) simply means 'evening flower', referring to the fact that it blooms at that time. Dame's rocket, sometimes known as dame's violet, is *Hesperis matronalis* because the flowers' heady scent is most strongly apparent in the evening air. The species name, incidentally, means 'of or befitting a married woman' and is related to the Roman festival of Matronalia, celebrating married women and childbirth and taking place on 1 March – though you'd be very unlikely in Britain to have the plant in flower by then. The specific epithet for evening has a change of initial letter and becomes *vespertinus*, as in vespers, the evening service of the Christian church. This, however, is a treacherous one. *Moraea vespertina* is reported to open its lemon-scented white flowers between 6 p.m. and 8 p.m., timekeeping that would have impressed Linnaeus, but *Asplenium vespertinum* is a fern and so does not of course have flowers at all. However, because the evening star appears in the west, *vespertinus* can also by association occasionally mean growing in the west, and the common name of this fern, native to California, is the western spleenwort.

Nightfall does not necessarily mean that plants go off duty, and the epithet *noctiflorus* (from *nox* (gen, *noctis*), the Latin for night) denotes those plants that come into their own after dark. The white flowers of the catchfly *Silene noctiflora* release their strong perfume into the night air in order to attract pollinating moths. The same goes for the night-flowering tobacco plant, *Nicotiana noctiflora*, and what I suppose might be called the night-time glory but is in fact known as the moonflower, *Ipomoea noctiflora* (syn. *I. alba*). There are two species of night-flowering jasmine, both taking different specific epithets relating to the dark. The first is *Cestrum nocturnum*, a large evergreen shrub native to the West Indies with greenish white tubular flowers, which can only be grown indoors in Britain. It is sometimes known as *dama de la noce*, or lady of the night, with all that the name might imply about its heady, darkness-penetrating scent. Equally fragrant is *Nyctanthes arbor-tristis* (already mentioned in an earlier chapter), as I discovered while walking back one evening after dinner along Rash Behari Avenue in South Calcutta. The intoxicating, overwhelming smell of jasmine hit me long before I came to florists' stalls hung about with glowing white garlands of these flowers, which are sometimes used in Hindu religious rituals and are, appropriately, the state flower of West Bengal. The plant's species name is discussed in 'Anatomy', but the genus name is from the Greek for 'night-flowering'.

Though not related to a particular season or time of day, *umbrosus* (from the Latin *umbra*, a shadow) is a useful specific epithet for shade-loving plants. Of course many plants that thrive out of direct sunlight have been labelled by botanists for other features, but among the species that advertise their preference for those otherwise difficult areas of the garden is *Saxifraga umbrosa*, which although native to the Spanish Pyrenees has been declared the 'true London Pride' by the RHS. Others insist that London Pride is the hybrid *Saxifraga* × *urbium* ('of cities'), which colonised bombsites during the Blitz and so inspired Noël Coward's morale-boosting 1941 song. Alice Coats points out that the hybrid saxifrage produces no seed, which means it would not have spread through the ruined capital in the way the species might have done; but in any case the common name dates back to the seventeenth century, before which a variety of sweet william (*Dianthus barbatus*) was known as Pride of London.

Sweet william, *Dianthus barbatus*

Equally important when planning a garden is the size and shape a plant will take. Some epithets related to size can at first sight be misleading, referring to a plant's flowers rather than its height or spread. *Senecio minutus* is indeed a very small species of ragwort, whereas the Mexican marigold *Tagetes minuta* can grow up to two metres tall, but has flowers that are not only pale yellow, but much smaller than most other species in the genus. Although the latter plant is useful for clearing ground of perennial and invasive weeds, neither is likely to appeal greatly to the average gardener. The same might be said of the 'very small' Bolivian milk vetch, *Astragalus minutissimus*, though other species such as *A. chlorostachys* ('green-spiked') and the purple *A. danicus* ('Danish') are grown as ornamental plants. There is a small species of red-flowered grevillea, native to Australia but named only in 1962, called *Grevillea diminuta*, but the epithet *diminutus* is more widely used to name bacteria and tapeworms. More common diminutive specific epithets are *pygmaeus, nanus, pumilus* or *pumilio*, and *parvus*. The last of these is simply the Latin word for 'small', and although not often used on its own supplies the frequently used prefix *parvi-* (see below). The others all mean 'dwarf', derived from words in their original languages applied to very small people. The two Latin words for a dwarf are *nanus* and *pumilio*, but *pygmaeus* has a rather more particular etymology. The Greek word *pygmaios* was applied to a mythical African race, but this name was itself derived from *pygme*, a unit of measurement based on the distance between the elbow and the knuckle of an adult human, rather as a cubit was based on the length of a forearm. All these epithets are widely used to describe small or dwarf forms of plants. The silvery dwarf harebell, *Edraianthus pumilio*, is a rockery plant native to Dalmatia much recommended by Reginald Farrer; and while the dwarf mountain pine is *Pinus mugo*, its specific epithet being derived from the tree's name in the Italian vernacular, the most often grown belong in a Pumilio group. Whereas it is evident why the dwarf bearded iris, or pygmy iris, is *Iris pumila*, and how the dwarf globeflower *Trollius pumilus* got its name, people have wondered why the apple should be called *Malus pumila*. Although apple trees can grow to seven metres, this name has been firmly designated the correct one, the apparently more sensible alternative of *M. domestica* having been rejected. The name was first used by Philip Miller in his *Gardener's Dictionary* of 1768, where he gives three kinds of apple tree, *Malus sylvestris*, the wild or crab apple, *M. coronaria*, the Wild Crab of Virginia, and *Malus pumila* 'which is rather a shrub than a tree, commonly called Paradise Apple'. It was this last, in the form of the 'Dutch Paradise Apple', that was 'much cultivated in nurseries for grafting Apples upon [and] generally preferred for planting espaliers or dwarfs, being easily kept within the compass usually allotted to these trees'.

Less controversial are the numerous dwarf plants labelled *nanus*, such as the dwarf birch, *Betula nana*, which grows only a metre high, and *Gladiolus nanus*, which can remain upright without the awkward and unattractive staking that makes other, larger species frankly unsuitable for growing anywhere but in a regimented bed for cutting flowers (if then). Among smaller shrubs are the dwarf spindle, *Euonymus nanus*, and a good number of 'Nanus', 'Nana' and 'Nanum' cultivars, all of them advertised as suitable for smaller gardens and certainly taking up a good deal less room than their larger relations. Being pygmy does not necessarily mean a plant lacks vigour. A kind nurserywoman once gave me a small plant of Corsican borage, *Borago pygmaea*, which she dug up from the gravel surrounding her sales area because she had none of the plant for sale. Years later, it still pops up all over my garden, its bristling little seedlings easy to spot and remove, but I always keep some because of their tiny china-blue flowers, the whole plant being a good deal less coarse than common borage, *B. officinalis*. The epithet *humilis* is also occasionally used of dwarf plants, and is a Latin word meaning low or lowly, derived from *humus*, which as all gardeners will know means the earth, soil or ground. The word 'humble' has the same etymology, and both this and *humilis* have the sense of someone or something

holding itself close to the ground. Compared with other jas-
mines, for example, the yellow-flowered *Jasminum humile* is a
low-growing species, while *Sarcococca hookeriana* var. *humilis*
is a dwarf variety of the Himalayan sweet box.

One last specific epithet denoting smaller species is *minor*
(neuter form *minus*), often used to differentiate plants from
major/majus, meaning larger. Such plants are often labelled
'lesser' and 'greater' in their English names, as with the lesser
and greater periwinkle, *Vinca minor* and *Vinca major*. Plants
are not always paired like this, however. While the familiar
garden nasturtium we all grew as children is *Tropaeolum
majus*, and the dwarf one *T. minus*, *Hosta minor* exists without a corresponding greater or *major* species. Similarly, the
beautiful white umbellifer *Ammi majus*, and the common or
garden snapdragon *Antirrhinum majus*, do not have accepted
minus counterparts. To be strictly accurate, there was a dwarf
snapdragon which Linnaeus called *Antirrhinum minus*, but it
has been renamed *Chaenorhinum minus*, which means 'little
gaping nose', and is a wild flower that looks better in the countryside than it would in a flowerbed.

There appear to be rather more specific epithets for plants
that grow big or tall. Tall plants can be merely *altus*, *elatus*
and *excelsus*, but get taller with *elatior* and *excelsior*, until they
are *altissimus*, the tallest of a species. As so often with species
names, one needs to be alert to other possible meanings.
Sedum altum is indeed an unusually tall and upright species
of this usually lax plant, and *Euphorbia alta* is known as the
giant spurge; but it would be hard to imagine a tall houseleek, and so the species name of *Sempervivum altum* refers to
the fact that it grows high in the Caucasus Mountains. The
Aralia genus includes plants that are both smallish border
specimens and large trees, *Aralia elata* coming into the latter

category, achieving heights of up to ten metres. Similarly, one of the common names for *Arrhenatherum elatius* is tall oat grass, while the oxlip, which does indeed look like a common primrose reaching for the sky, is *Primula elatior*. The Chilean *Lobelia excelsa* produces spikes of orange-red flowers that can be between one and a half and two metres tall, while the English ash tree, *Fraxinus excelsior*, can grow as tall as forty-three metres, though it more usually reaches a little over half that height. Compared to the ash, the tree of heaven, *Ailanthus altissima*, may seem small, but it gets both its common and Latin names from its rapid growth, shooting heavenwards to achieve a not inconsiderable height of fifteen metres in as little as twenty-five years. Its genus name, incidentally, also suggests its vertical reach, since it is a Latinised version of its vernacular name in the Spice Islands (the Maluku archipelago), *ailanto*, meaning 'sky tree'.

The epithet *procerus* carries with it the notion of nobility as well as height, two characteristics often linked in the ancient world, where *procerus* was merely another Latin word meaning high or tall, but *procer* meant a great or noble person. One school of thought suggests that *Abies*, the name given to the fir genus in classical times, derives from the Latin verb *abere*, meaning to depart, because firs leave the ground to soar to a great height in the sky. Particularly tall is *Abies procera*, which can reach a height of seventy metres and has the common name of the noble fir. Another epithet that can mean either large or noble or both is *nobilis*. It might be hard to discern anything particularly noble about chamomile, and its genus name is the Greek for 'ground-apple', a reference to the fact that it grows at ground level and has a scent reminiscent of apples. As Gerard notes, however, Roman chamomile, *Chamaemelum nobile*, is 'stiffer and stronger than any of the others, by reason whereof it standeth more upright, and doth not creepe vpon the earth as the others doe'. The nobility of the bay tree is not in dispute: its leaves were used to wreathe the heads of victors and emperors in the ancient world, and are still a symbol of honour, whether you are crowned with laurels or merely resting on them. Many gardeners grow bay trees in pots, sometimes clipped into lollipops, but in the wild *Laurus nobilis* can reach a height of eighteen metres and so earns its species name in that way too. Roman nobility leads us to the emperor's epithet, *imperialis*, which means showy or stately, characteristics also connected with size. The crown imperial, *Fritillaria imperialis*, certainly lives up to its name, sending up a tall stalk hung about with large burnt-orange bells surmounted by a topknot of leaves like the crest of feathers on a governor-general's cer-

emonial hat. It came to Europe in 1576, and although it was first brought from Persia and known as 'the Persian lily', Gerard wrote that his own specimen came from Constantinople, a frequent source of bulbs in England at that time. Equally spectacular, and absolutely huge, is the tree dahlia, *Dahlia imperialis*, an introduction of 1868 that can reach a height of some nine metres and in autumn has bright-lilac single flowers with vivid ginger bosses.

Epithets for plants that are just big rather than tall range from *magnus* and *grandis* to *giganteus*, the meanings of which are easy

The crown imperial,
Fritillaria imperialis

enough to decode. The Star of Bethlehem *Ornithogalum magnum* is differentiated from other bulbs in the genus by its tall spikes of white flowers, large enough perhaps to guide the Three Wise Men. The epithet *grandis* can also mean showy, and the red-flowered *Eriogonum grande*, unlike other buckwheats, is showy enough to earn its place in a garden. Gigantism in plants tends to be comparative, although there are some species that are massive by any standards, such as the giant redwood *Sequoiadendron giganteum*. The Himalayan lily *Cardiocrinum giganteum* towers above other lilies, and indeed above anyone who manages to grow it; and, although not on the same scale, the beautifully light-catching golden oat, *Stipa gigantea*, is much larger than other grasses in the genus. *Gigas* is simply the Greek word for a giant, adopted into Latin, and is familiar from the majestic purple *Angelica gigas*. A favourite among gardeners who like architectural and dark-hued plants, it is certainly dramatic but doesn't in fact grow any larger than common angelica, *A. archangelica*.

Plants labelled with the epithet *maximus* can be the largest of their species, such as *Hepatica maxima*, the biggest of the (admittedly small) liverworts, or may instead have some other feature, such as flowers or fruit, larger than those of others in the genus. *Astrantia maxima*, for example, has bigger pink flowers than *A. major*, and the cobnut, *Corylus maxima*, has larger nuts than the hazelnut, *C. avellana*, which gets its species name from the ancient city of Avella Vecchia in the foothills of the Apennines, where Pliny had recorded the tree growing wild. Plant species can also be graded by size, as the quaking grass *Briza* is, from *Briza minor*, which is both the smallest and has the tiniest flowers, via the medium-sized *B. media*, to *B. maxima*, which can grow to around twice the height of its smallest relative and has by far the largest flowers.

As well as the size a plant will reach when mature, information about its habit is also useful to gardeners. There are a vast number of epithets covering this aspect, some of which will be either familiar or easily guessable. Others seem less obvious. Because it is not reliably hardy, I grow *Salvia patens* in a pot, which does not allow me to see that, given space, it would spread, which is what *patens* means, from the Latin *patere*, to lie open or extend. Another epithet for spreading is *procumbens*, from *procumbere*, meaning to bend forward or prostrate oneself, and is used of plants such as *Juniperus procumbens*, which rather than growing vertically spreads its trailing branches over a wide area. Sounding rather similar is *procurrens*, but this is from the Latin verb *procurrere*, to run or rush forward, as the lovely soft-magenta *Geranium procurrens* does, sending out new plants on long runners to root in new ground. *Stoloniferus* (from the Latin *stolo* (gen. *stolonis*) = a shoot + *ferre* = to bear) refers to the same habit, and is the way the stonecrop *Sedum stoloniferum* propagates itself. The species names *repens* and *reptans* come from the Latin verbs *repere* and *reptare*, both of which mean to creep. White clover, which is welcome when it creeps across a meadow but less so when it invades a lawn, is *Trifolium repens*, and bugle, which is grown in a large number of garden cultivars and makes an attractive groundcover plant, sending up triumphant little turrets of blue, pink or purple flowers as it conquers a territory, is *Ajuga reptans*. Less familiar is *divaricatus*, which also means spreading widely or even straggling, from the Latin *divaricare*, to stretch apart; but when I look at how the dark-stemmed, low-growing aster *Eurybia divaricata* colonises the bed in which I planted it, I see how accurate this epithet really is.

Plants that grow upright have several possible species names. The most obvious one is *erectus*, as in the so-called

African marigold *Tagetes erecta*, which grows tall and is in fact a native of Mexico. Other epithets are derived from Latin words that add a moral dimension to being 'upright'. *Rectus* is the Latin word for right or straight, and is the origin of such English words as rectify and rectitude. It is familiar from *Clematis recta*, a species that unlike its climbing cousins is an upright shrub. (The species name of one of those cousins, *Clematis viticella*, simply means 'little vine'.) *Strictus* has an element of upright containment, derived from *stringere*, a Latin verb meaning 'tighten'. *Verbena stricta* could be described as a fine upstanding species with rigid spikes of flowers, unlike the idly sprawling, brightly coloured plants for tubs and hanging baskets most of us still call verbenas but in fact belong to the related *Globularia* genus. You might think *columnaris* would mean straight up and down like a column, but it usually describes a species with a habit more upright than others in the genus, such as the New Caledonian pine, *Araucaria columnaris*. It is more often used as a cultivar name, as in the Norway maple, *Acer platanoides* 'Columnare', or the Lawson's cypress *Chamaecyparis lawsoniana* 'Columnaris', which planted in rows is more convincingly columnar than most trees. *Fastigiatus* suggests narrowly upright, tapering to a point, from the Latin *fastigium*, which in origin means a roof gable; once again the epithet is employed chiefly as a cultivar name. *Ilex crenata* 'Fastigiata' is a columnar form of the Japanese holly, while the Irish yew, *Taxus baccata* 'Fastigiata', is thought to be a mutant form of the common yew, discovered in County Fermanagh in 1780 and subsequently brought into cultivation.

There are several specific epithets for plants that twist or droop or otherwise fail to hold themselves in a properly vertical position. *Allium obliquum* has stems well out of true and is unkindly known as the lopsided allium, while trees that take a somewhat twisted or tortured form, often because they grow on coasts, are sometimes allotted the specific epithet *contortus*. The straight trunk of the lodgepole pine, as its name suggests, was used in its native America for building, but it gets its botanical name, *Pinus contorta*, from the fact that when it remains a shrub it is often stunted and deformed. The corkscrew hazel, sometimes known as Harry Lauder's walking stick because the Scottish music-hall artist carried a twisted cromach (not in fact made of hazel), is not a species but a cultivar, *Corylus avellana* 'Contorta'. And although some nurseries offer '*Salix contorta*', no such tree exits; the correct name for the twisted willow, another cultivar, is *Salix matsudana* 'Tortuosa'. A further twisty epithet is *flexuosus*, though it is not always apparent why it is used. The RHS concedes that the sinuate bamboo, *Phyllostachys flexuosa*, has canes that are 'slightly wavy', but there doesn't seem to be anything more particularly wavy about *Corydalis flexuosa* or *Helenium flexuosum* than other species.

Some 'weeping' plants are described elsewhere, but the epithet *pendula*, from the Latin *pendere*, to hang, is commonly used of both trees and herbaceous plants. The silver birch is *Betula pendula* because the branches hang down, and the angel's fishing rod *Dierama pendulum* owes its species name to the way the flower-heads are gracefully suspended from its wiry stems. Also denoting a weeping habit is *suspensus* (similarly derived but with the addition of the Latin prefix *sub-*, meaning under or below), as in the weeping forsythia, *F. suspensa*. You may feel that the most famous of all mourning trees, the weeping willow, should come in here, but its botanical name is *Salix babylonica*. The tree is in fact native to China rather than Babylon, and Linnaeus, who named it, got his specimen from a garden in the Netherlands. I thought at first that the explanation for this species name is that Linnaeus was reminded by its habit of the Hanging Gardens of Babylon. It has been pointed out to me, however, that he was more likely to have had in mind the famous opening of Psalm 137: 'By the rivers of Babylon, there we sat down, yea, we wept, when we remembered Zion,' which continues: 'We hanged our harps upon the willows in the midst thereof.'

When it comes to climbing species, the most usual specific epithet is *scandens*, from the Latin *scandere*, to climb. It is sometimes used to differentiate a climbing species from its earthbound relatives, as in *Dicentra scandens*, which not only scrambles vigorously to a height of some four metres but has 'bleeding heart' flowers that are a bright canary yellow. This was not good enough for the taxonomists, who have now shifted

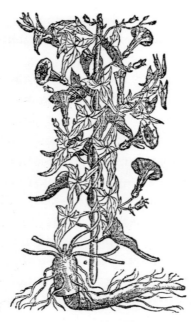

'Syrian Scammonie',
Convolvulus scammonia

it to the largely unfamiliar *Dactylicapnos* genus. Even more energetic in scaling frames, walls, and more or less anything else in its path is *Cobaea scandens*, known because of the structure of its purple and green flowers as the cup-and-saucer plant. Although tender, it is capable of scrambling to a height of six metres in one growing season. The *Convolvulus* genus takes its name from *convolvere*, meaning to roll round or intertwine, and the morning glory is *Ipomoea* from the Greek *ips* = worm + *homoios* = similar.

Other plants that twist and twine their way around supports as they reach for the empyrean have the species name *volubilis*, the Latin word for spinning or rotating. Given the right conditions, the Australian wax jasmine, *Jasminum volubile*, can send out enormously long curling stems, and haul itself up nine metres. The climbing aconite, *Aconitum volubile*, has the same dark blue-violet flowers as many of the herbaceous species of monkshood, and will usually be found in nurseries under this name. We are, however, sternly warned by the RHS that the name has been 'misapplied' and that the plant should instead be called *Aconitum hemsleyanum* after a Kew botanist called William Botting Hemsley, co-author of such works as *An Enumeration of All the Plants Known from China Proper, Formosa, Hainan, Corea, the Luchu Archipelago, and the Island of Hong Kong* (1887).

Other species names tell us how far a plant resembles a shrub, from *frutescens* and *suffruticosus*, meaning somewhat shrubby or a subshrub, to *fruticosus* and *fruticans*, meaning properly shrub-like. All these words are derived from the Latin *frutex* (gen. *fruticis*), meaning a bush (or, for some reason, a blockhead). Epithets taken from *arbor*, the Latin word for tree, are used the more woody a plant gets: the tree mallow is *Lavatera arborea* (syn. *Malva arborea*), and the tree wormwood, most often grown as the feathery 'Powis Castle' cultivar, is *Artemisia arborescens*. Plants can be *ramosus* (branched) or *ramulosus* (twiggy), or simply *dumosus* (bushy, from the Latin for briar), as in the bushy aster *Symphyotrichum dumosum*; or they can be *compactus* (compact) and *concinnus* (neat), as in the tidy little lupin *Lupinus concinnus*.

We need not spend too much time on leaf types, but two words are essential to know because they often form a component of specific epithets. Not surprisingly, these are the

Latin and Greek words for leaf, *folium* and *phyllon* (Latinised as *phyllum*). For example, *grandifolius* and *macrophyllus* both mean having large leaves. *Grandis* is the Latin for large, and, as previously explained, in botanical Latin the Greek-derived prefix *macro-* usually means large. (Both *grandi-* and *macro-* or *macr-* can be attached to describe other large features of a plant: *grandiflorus*, with large flowers, *macracanthus*, with large spines.) Although it has some of the prettiest flowers of all the St John's worts, *Hypericum grandifolium* is named because of the distinguishingly large size of its leaves. On the other hand, one would have thought that what separates out *Muscari grandifolium* from other grape hyacinths is its bicolour flowers of china blue and very dark purple, but botanists preferred to look at its leaves. It should be fairly evident that the opposite of these size epithets are *parvifolius* and *microphyllus*, once again generally used to mark out plants that have smaller leaves than those of other species within a genus. The leaves of the Chinese elm, *Ulmus parvifolia*, which was introduced to Britain at the end of the eighteenth century, are much smaller than those of the almost late and very much lamented English elm, *U. minor* var. *vulgaris*. The Greek-derived epithet *microphyllus* is much more often used and is familiar from the Japanese box, *Buxus microphylla*, the dense habit of which makes it a popular alternative to *B. sempervirens* for edging flowerbeds and making knot gardens. Plants with smaller leaves than usual are sometimes aesthetically desirable, and I'd argue that this was the case with *Fuchsia microphylla*, introduced in the 1820s, with leaves both smaller and prettier than most other species, a further advantage being that it is rather more hardy.

Numerous other prefixes can be attached to *-folius* and *-phyllus* for other kinds of leaf: *longifolius*, very obviously with long leaves; *stenophyllus*, less recognisably, with slender leaves, from the Greek word *stenos*, meaning narrow. It will be very evident when looking at different species of lungwort in a nursery how *Pulmonaria longifolia* got its name. Equally the elegant, arching *Polygonatum stenophyllum* has far narrower leaves than the ordinary Solomon's seal. The epithet *angustifolius* means the same thing, *angustus* being the Latin word for narrow. It will be familiar from common or English lavender, *Lavandula angustifolia*, and rosebay willowherb, *Epilobium angus-*

Solomon's seal, *Polygonatum*

tifolium, the white variety of which, var. *album*, has recently become a fashionable border plant. The botanical opposite of *angustifolius* is *latifolius*, from the Latin *latus*, meaning broad, and this is the specific epithet that helps differentiate English and Portuguese lavender, *L. latifolia*, which indeed has stubbier and broader leaves. Alternative epithets for narrow or very fine leaves include *tenuifolius*, from *tenuis*, the Latin word for slender. The pale flowers of the alpine *Gypsophila tenuifolia* hover above a mound of grass-like leaves, and *Paeonia tenuifolia* carries its red flowers above a feathery mass of bright green foliage very different from the leaves of other peonies. The Greek-derived alternatives to *latifolia* are *platyphyllus* and *platyphyllos*, used to identify plants as different in overall size as the unusually broad-leaved snowdrop, *Galanthus platyphyllus*, and the more familiar large-leaved lime, *Tilia platyphyllos*.

The meaning of *rotundifolius*, from the Latin *rotundus*, will be apparent from similarly derived English words for people who are stout (rotund) and a building with a circular ground-plan (rotunda). That favourite of hanging baskets *Pellaea rotundifolia* is known as the button fern because of the round leaflets of its fronds. The delicate violet-coloured flowers nodding on top of wiry stems may be what first strikes most of us about the harebell, but for botanists it is the leaves that define the plant as *Campanula rotundifolia*.

Some specific epithets can refer both to leaves and to petals. Those that taper to a point are called acuminate, from the Latin *acuere*, to sharpen. The leaves of the emphatically named evergreen honeysuckle *Lonicera acuminata* var. *acuminata*, for example, are very different in shape from the rounded ones of most species; but the best example of a plant taking this specific epithet is the miniature tulip *Tulipa acuminata*, because the petals of its fiercely burning flowers are yellow at the base but bright scarlet at their highly attenuated tips. And indeed *attenuatus* is another epithet for things that narrow to a point, most sharply in such succulents as *Agave attenuata* and the zebra plant, *Haworthia attenuata*, but also in the taperleaf penstemon, *Penstemon attenuatus* var. *attenuatus*. There is another more particular word for flowers with pointed petals, *acutipetalus*, and the prefixes *acu-* and *acuti-* are used in species names to denote anything sharply pointed – *acutifolia* (leaf), *acutilobus* (leaf lobes), *acutissimus* (very acutely pointed) – or simply prickly. Indeed *acus* is the Latin word for a needle. As so often with such words, some of them look very alike. The hard shield-fern and the purple-flowered Waldo rock cress are clearly very different in appearance, but their species names are uncomfortably close: *Polystichum aculeatum* (meaning prickly) and *Arabis aculeolata* (meaning with small prickles). Such differences matter more to botanists than to gardeners, who are unlikely to be prickled very seriously by either the bristly tips of the former's leaves or the stiff white hairs of the latter's flower-stems. Other epithets for pointiness include *subulatus*, taken from *subula*, the Latin word for an awl, and more or less synonymous with *acuminatus*. Certainly the narrow little leaves of that rockery favourite the creeping phlox, *Phlox subulata*, might be described as needle-like. One final, potentially misleading, epithet is *pungens*, which at first glance looks as if it might refer to a plant's smell. This seems a distinct possibility when we think of the pleasantly resinous odour of the blue spruce, *Abies pungens*. The Latin word from which we get both this epithet and the English word pungent is *pungere*, meaning to prick: the epithet in fact refers to the needles of the spruce, on which you might well prick a finger, whereas something that is pungent will 'prick' our nostrils (which is why we talk about a 'sharp' or 'penetrating' odour).

Just as sharp, but larger than awls and needles, is the sword, one of several types of weapon evoked by botanical names. We have already mentioned *ensis*, and among plants named for their sword-like foliage are the Japanese iris *Iris ensata*, and the red-hot poker *Kniphofia ensifolia*, the leaves of which are not only sword-shaped but also sharp-edged. Another and more familiar Latin word for sword is *gladius*, hence gladiator and *Gladiolus*, the name once again referring to the leaves that give the plant its common name of sword lily. Other weapons found on ancient battlefields include the spear, the arrow and the shield, and each of these has provided specific epithets which will sound familiar to anyone who has slogged through the kind of Latin lessons that seemed mostly to involve translating sentences about Caesar's wars. The Latin for a spear is *hasta*, giving us *hastatus* for plants with foliage or flower-spikes resembling spears. The flower-spikes of *Verbena hastata* do look a little like an army of spears bristling before battle, especially because the individual flowers open from the bottom of the spike, leaving the tip looking sharp; but such cultivar names as 'Blue Spires' somewhat confuse the image. Other epithets for spear-like are taken from a second Latin word for this weapon, *lancea*, as in the narrow-leafed plantain, *Plantago lanceolata*, or the similarly slender-leafed hosta, *Hosta lancifolia*.

Despite the best efforts of botanists, other plants described by them as spear-like have often acquired common names taken from other weapons, perhaps excusably so in the arrowleaf dock, *Rumex hastatus* (the tips of arrows and spears often being very similar), less so in the silver sword, *Philodendron hastatum*. The Latin word for arrow is *sagitta*, giving us Sagittarius, the astrological sign of the Archer. The water-plant *Sagittaria sagittifolia*, a British native with purple-centred white flowers, is unsurprisingly called arrowhead because of its distinctively shaped leaves. The leaves of the ivy *Hedera helix* 'Sagittifolia' suggest a rather more complex arrow that, once having penetrated a body, would be hard to pull out because it has several lower lobes.

The specific epithet *peltatus* is usually defined as shield-shaped, and indeed comes from a Greek word for a shield, *pelte*. It is confusing that the most familiar species with this name is *Pelargonium peltatum*, which is commonly known as the *ivy-leaf* geranium. In fact peltate is a botanical term meaning that the petiole (leaf-stalk) is attached to the centre of the leaf rather than to the lower margin, presumably derived from the fact that this is where the handle of a shield would be attached. The example usually given of this kind of leaf is that of the sacred lotus *Nelumbo nucifera*, although a plant you are more likely to see growing in British gardens with a bog area, *Darmera peltata* (fancifully called Indian rhubarb), would be another example.

Plants are also named because of the pattern or marking of the margins of their leaves. Leaf margins can be *undulatus* (wavy), *laciniatus* (deeply cut), or *fimbriatus* (fringed). There are several words for different kinds of tooth-like leaf edges, among them *dentatus* (toothed), *denticulatus* (fine-toothed), *crenatus* (round-toothed) and *serratus* (saw-toothed). Some plants with margins that are marked in a different colour to the rest of the leaf have the species or cultivar names *marginatus* or (less often) *marginalis*. *Euphorbia marginata*, for example, has broad and bold white edges to its grey-green leaves, while *Crassula pellucida* subsp. *marginalis* f. *rubra* is a brightly coloured succulent with purplish leaves bordered with red. More often *marginata* is combined with a colour to create cultivar names, such as gold-edged *Daphne odora* 'Aureomarginata'. Although the fern *Dryopteris marginalis* is sometimes cited as an example of a plant with a distinct margin, its name in fact refers to the clusters of brown sporangia neatly arranged on the underside edges of its fronds.

Other epithets referring to markings include *zonalis* (from *zona*, the Latin word for a belt), meaning leaves with irregular bands of different colour, familiar from *Pelargonium zonale* in its many cultivars. *Striatus* means striped: either the leaves, such as the cream and grey-green foliage of *Sisyrinchium striatum*, or the flowers, such as those of *Geranium sanguineum* var. *striatum*, which are a pale pink marked with darker pink veins. *Pictus* literally means painted and is used to define the beautifully marbled *Arum italicum* 'Pictum' and the spectacularly yellow-mottled *Abutilon pictum* 'Thompsonii'. Smaller markings on leaves or flowers provide the species names *maculatus* and *punctatus*. The Latin word *macula* can mean a spot, but also a stain, hence the English word 'immaculate', meaning pure and unblemished. The spotted deadnettle, grown in gardens in a number of variously splotch-leaved cultivars, is *Lamium maculatum*, and the heath spotted-orchid *Dactylorhiza maculata* has both boldly marked leaves and well-speckled flowers. The spotted bee-balm, *Monarda punctata*, has pale yellow flowers mottled with purple, and the flowers of the true *Campanula punctata* are lightly spattered, though this distinction is sometimes lost in cultivars. The particular species name for plants with speckled flowers is *guttatus*, from the Latin *gutta*, meaning a spot, drop or speck. This is seen in the monkey flower *Mimulus guttatus*, which has an ominously consumptive sprinkling of blood-red spots in its yellow throat. A species name to do with markings that puzzles many gardeners is that of *Iris reticulata*. Reticule is a word one doesn't often come across these days, but was once widely used for a woman's small handbag usually made of cloth, its name derived from *reticulus*, the Latin for a little net or a bag made of mesh. In botany *reticulatus* denotes a plant that is 'netted', as in the strongly veined petals of *Camellia reticulata*, the similarly veined leaves of the blueberry ash, *Elaeocarpus reticulatus*, or the distinctively marked fruit of the custard apple, *Annona reticulata*. And the iris? Gardeners need to remember that botanists sometimes ignore the most striking feature of a plant, such as this one's bright spring flowers, and the species name instead refers to the netted sheath of the bulb.

As with leaves, when it comes to flowers, the two important words often used with prefixes in the creation of specific epithets are already familiar: *-florus* and the Greek-derived *-anthus*. The latter will be recognised from polyanthus, which

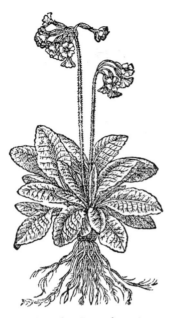

Cowslip, *Primula veris*

means the same as *multiflorus*, 'many-flowered', but is not in fact the botanical name for that hardy plant whose flowers of many colours herald the spring. The origins of 'polyanths' (as many gardeners anglicise them) are lost in time, but it seems they emerged as a cross between coloured varieties of the primrose and the cowslip and had the botanical name *Primula × polyantha*. The word polyanthus is recorded by the diarist and gardener John Evelyn in 1687, and by 1728, when the 'Spring' section of James Thomson's poem *The Seasons* was first published, this plant was clearly differentiated from the woodland primrose:

> Fair-handed spring unbosoms every grace;
> Throws out the snowdrop and the crocus first;
> The daisy, primrose, violet darkly blue,
> And polyanthus of unnumber'd dyes . . .

By 1774 James Gordon would describe them in his *Planters, Florists, and Gardeners Pocket Dictionary* as 'the most esteemed' of all the primulas, and noted that they were 'subdivided into three other classes, viz., the common Polyanthus, the Hose-in-hose, and the Pantaloon', the latter two being double forms. The epithet *polyanthus* is, however, used for such floriferous plants as the pink jasmine, *Jasminum polyanthum*.

The form taken by many species names that define flowers is similar to that of leaves, so that both *grandiflorus* and *macranthus* mean with large flowers, as in the yellow foxglove, *Digitalis grandiflora*, and *Exochorda* × *macrantha*, which (particularly in such cultivars as 'The Bride') has such cascades of large white flowers in spring that very little foliage remains visible. Plants that flower in such profusion might also have the species names *floridus* (not to be confused with *floridanus*, meaning from Florida) and *floribundus*, the latter familiar from 'Floribunda' roses, which are not a species but a group of modern hybrids that have *Rosa multiflora* (syn. *R. polyantha*) among their forebears. A similar epithet is *densiflorus*, fairly obviously meaning densely flowered, particularly noticeable in the packed yellow flower-spikes of the mullein *Verbascum densiflorum* or the glowing orange pokers of the ginger lily *Hedychium densiflorum*. The epithet *nudiflorus* seems an odd one, for what constitutes a naked flower? However, the Latin verb *nudare*, to bare, also has the sense of to expose; so in botany *nudiflorus* is used of plants on which the flowers appear before the leaves, such as the yellow winter jasmine *Jasminum nudiflorum*, and are therefore more fully exposed without the backing of foliage. The species name *pauciflorus* means with few flowers, which is hardly enticing but can also be misleading, and is for this reason worth mentioning. One can see that, compared with other species of red-hot pokers, the bright yellow *Kniphofia pauciflora* has fewer flowers on its spikes and that they are more spread out; but the whole point of *Corylopsis pauciflora* is that it has a mass of very fragrant flowers that appear in the winter, so the name is a mystery. Similarly *Polemonium pauciflorum* is an unusual and very free-flowering species of Jacob's ladder with delicate trumpets of murky yellow striped with dusky red that are quite unlike the flowers of others in the genus. The explanation for the name in this case may be that the flowers of other species, differently shaped and usually blue or purple, grow in denser clusters.

The way flowers are arranged on the plant are described in such species names as *paniculatus*, ultimately derived from the Latin *panus*, an ear of millet, and so meaning in loose clusters or panicles. Hence the popular, late-flowering *Phlox paniculata*, as well as *Hydrangea paniculata*, which has flowers in much less compact heads than those of the lace-cap *H. macrophylla*. Plants with flowers held in dense clusters are denoted by *glomeratus*, from the Latin *glomerare*, meaning to mass together, beautifully exemplified by *Campanula glomerata*, while *capitatus* means flowers forming a dense head, as do the dark violet ones of the Himalayan primrose *Primula capitata*. *Racemus* is the Latin word for a bunch, particularly of grapes, which gives us the epithet *racemosus* for plants with clusters of flowers growing in a similar manner along a stem, as in the false Solomon's seal, *Smilacina racemosa*, or the catmint *Nepeta racemosa*, of which 'Walker's Low' is a well-known cultivar. *Umbella* is a Latin word for parasol, and so *umbellatus* denotes plants with flowers that grow from pedicels arranged like umbrella spokes to form a domed or flat shape. The candytuft *Iberis umbellata*, the flowering rush *Butomus umbellatus*, and the white flat-topped aster now burdened with the botanical name *Doellingeria umbellata* all take this form and name. *Spicatus* denotes plants with flowers arranged in spikes, such as the blazing star *Liatris spicata* and the tall speedwell, *Veronica spicata*. *Sphaerocephalus* or *sphaerocephalon* denotes round flower-heads, such as those of the ghostly white globe thistle *Echinops sphaerocephalus* 'Arctic Glow', and the maroon drumsticks of *Allium sphaerocephalon*. Derived from the globe rather than the sphere, *Buddleja globosa* has very different flowers from other species, not just because they are orange but because they are arranged in globes rather than panicles. Though perhaps the least known and grown of the buddleias, it was the first one to be introduced to Britain, in 1774, and was once marketed as 'the Globose Buddlebush', a name that makes the Latin one sound elegant. Plants with flowers that are bowed or nodding, or even drooping, sometimes have the species names *cernuus*, a Latin adjective meaning stooping forwards, or *nutans*, from the Latin verb *nutare*, to nod. *Allium cernuum* has lovely clusters of lilac flowers that hang down gracefully from their stalks, while the grain-like silvery brown flowers of the grass *Melica nutans* are elegantly suspended from their bowed stems.

Many species names for different forms and shapes of flowers, like those for leaves, are really the province of the botanist, and not particularly helpful to the gardener. A good number of plants, for example, have the specific epithet *involucratus*, which means that they have a circle of bracts around several flowers. This may help scientists to differentiate between species, but is not perhaps the first thing the gardener looks for when selecting a plant at the nursery. Other epithets sound less technical, but are not necessarily more useful. For instance, some plants with bell-shaped flowers are allotted the epithets *campaniflorus* or *campanulata* from the Latin *campana*, which also gives us the genus of bell-flowers *Campanula*. The difficulty for gardeners is that a great many plants have flowers that resemble bells. *Clematis campaniflora* does, for example, but then so do many other clematis. *Campanulata* is a diminutive and suggests little bells, quite rightly in the case of *Enkianthus campanulatus*, though it is hard to see why the flowers of *Agapanthus campanulatus* or *Rhododendron campanulatum* are more suggestive of bells than those of others in the genus.

That said, there are a couple of specific epithets relating to flowers that do provide gardeners with practical information. If you want to select species, forms, varieties and cultivars of plants with double flowers, you can look out for *plenus* (the Latin word for 'full', and the root of 'plentiful'), *pleniflorus* and *flore pleno*. A number of clematis with double or complex flowers have the cultivar name 'Plena', such as the white and green *C. florida* 'Plena' or *C.* 'Purpurea Plena Elegans', with flowers so doubled as to remind me of the many-layered rosettes that used to be given to champion cattle at the agricultural shows of my childhood. The first of these plants is now more correctly *Clematis florida* var. *flore-pleno* 'Plena', which seems to leave little doubt about the matter, and the same might be said of the double snowdrop *Galanthus nivalis* f. *pleniflorus* 'Flore Pleno', with its intriguing central muddle of green and white petals. The pink double hawthorn *Crataegus laevigata* 'Rosea Flore Pleno' adds colour to the cultivar name, as does the double meadow cranesbill *Geranium pratense* 'Plenum Violaceum'.

Another common epithet is *barbatus*, from the Latin *barba*, a beard. The flowers of sweet william, *Dianthus barbatus*, a favourite plant in Britain since its introduction in 1573, have hairs on the inside. Similarly, there are hairs on the lips of the tubular flowers of *Penstemon barbatus*, and in America penstemons are known as beardtongues. It may seem odd that the bearded iris, so called because of the sometimes different-coloured patches of hair on the standards, does not have the epithet *barbata*. Bearded irises are not, however, a species, but a group, and they are sometimes called pogon irises, *pogon* being the Greek word for a beard. (Why a genus of grasses should be named *Ophiopogon*, which translates as 'snake beard', remains a total mystery.) There is also a theory that the name of the *Verbascum* genus is a corruption of *barbascum* or bearded, referring to the downy leaves of such species as the greater mullein, *V. thapsus*, or perhaps to the plant's hairy stamens. Whatever the case, the name dates back to Pliny.

The much depleted *Aster* genus takes as its name the Greek word for a star. The one aster not redistributed to other genera, *Aster amellus*, was the first to be introduced to England, from Italy before 1596, and appears in the 1633 edition of Gerard's *Herball* as *Aster Italorum* or Italian Starrewoort. There may not be much of the star about this species, but the first one listed by Gerard, who calls it *Aster Atticus*, has flowers 'of a shining and glistering golden colour; and underneath about these floures grow fiue or six long leaues, sharpe pointed and rough, not much in shape unlike the fish called *Stella marina*'. The accompanying illustration does indeed show stalks bearing what look like starfish, but it is unclear what species of plant this really was, since *Aster atticus* is now deemed a synonym of *A. amellus*. Both image and description suggest the spiny starwort, *Asteriscus spinosus*, which is also a member of the aster family. Other plants with flowers that are variously deemed star-shaped go to *stella*, the Latin equivalent of *aster*, for the species name *stellatus*. *Magnolia stellata*, for example, has open white flowers with whirligig petals very different from those of other species, and the flowers of the campion *Silene stellata*

Gerard's 'Starrewoort, *Aster Atticus*'

are a pure white starburst of five jagged-edged petals. The species name of *Scabiosa stellata* refers not to its flowers, but to its extraordinary and highly decorative spherical seedheads, which look like some planet out of a science-fiction story.

Perhaps the most promising species name relating to flowers is *callianthus*, from the Greek *kallos* = beauty + *anthos* = flower. Beauty is, of course, in the eye of the beholder and so, unlike many other species names, can hardly be relied upon. I'd certainly argue that the fragrant, purple-splotched white flowers of *Gladiolus callianthus* are far more beautiful than those of most gladioli; but it would seem the botanists now disagree since they have renamed the plant *Gladiolus murielae*, after a no doubt deserving Mrs Muriel Erskine, whose husband collected a specimen in Ethiopia in the 1930s. The yellow flowers of *Berberis calliantha* are unarguably pretty, but are they any more so than those of other species? And how is one to decide between two species of *Desmodium*, one called *D. callianthum* and one *D. elegans*? The prefix *calli-* is widely used for other aspects of plants, from *callicarpus* (with beautiful berries) and *callizonus* (beautifully banded) to the unlikely *callistachyus*, which means beautifully spiked and is in fact justified by South American *Odontonema callistachyum*, the purple flower-spikes of which are indeed beautiful. The botanical name of the China aster, *Callistephus*, does at least refer to the flower's shape, *stephos* being a Greek word for crown. And let's not forget the bottle-brush genus *Callistemon*, its name meaning 'beautiful stamens'. Ordinarily, stamens attract the interest of scientists and hybridisers rather than gardeners, but here is an example of a plant which would look nothing without these brightly bristling reproductive organs.

These 'beautiful' species names are not the only ones that do not really mean a great deal. We all know *Anemone blanda*, but what precisely is so distinctively 'pleasing' or 'charming' about it? And is *Rudbeckia gloriosa* really more glorious than other coneflowers? I love agapanthuses, and grow several species of them, but am still not sure why its name means 'love-flower' (from the Greek *agape* = love + *anthos* = flower). And couldn't Pliny have come up with a better name for the daisy than *Bellis*, derived from *bellus*, the Latin for pretty? It could, I suppose, be argued that the true bleeding heart, with its contrasting white drips apparently oozing from its magenta lockets, earns its name of *Dicentra spectabilis*, and that *Lilium regale* is indeed the most regal of the lilies. But even these generally flattering names are not always reliable. *Euonymus*, for example, is a Greek word meaning of good name or reputation, but it was given to this genus of shrubs and trees as a dark joke, since they can be poisonous to livestock. Finally, it is worth mentioning that *speciosus*, though it has the same Latin root, does not mark out a plant as the original species but merely means that it is showy or handsome, *species* being a Latin word for appearance, but also for display, splendour and beauty. Generally speaking, however, such epithets show early botanists perhaps rightly astounded at their discoveries and reaching for a name that conveys their delight. And looking round at the vast array of plants now available to us, who can blame them?

Word Lists

TIME AND SEASON

aestivalis, aestivus = of the summer

autumnalis = of the autumn

brumalis = of the winter

hesperus = of the evening

hibernus = winter-flowering or green in winter

hiemalis = winter-flowering

majalis = may-flowering

meridianus, meridionalis = flowering at midday

noctiflorus = night-flowering

nocturnus = of the night

praecox, praevernus = very early, before spring

primula = first-flowering

serotinus = late-flowering, autumnal

vernus, vernalis, veris = of the spring, spring-flowering

vesperis, vespertinus = of the evening (but also western)

HABIT

amplexi- (prefix) = clinging

anfractuosus = bent, twisted

angularis, angulatus = angular

arborescens, arboreus = woody, treelike

bistortus = twisted

caespitosus = clump-forming

columnaris = columnar

compactus = compact

concinnus, concinnatus = neat

contortus = twisted

decumbens = prostrate

with upturned ends

dependens = hanging

dilatatus = spreading

divaricatus = spreading, straggling

dumosus = bushy

effusus = spreading

erectus = upright

fastigiatus = upright and narrow or narrowing to a point

flexuosus = tortuous or wavy
frutescens = somewhat shrubby
fruticosus, fruticans = shrubby
gracilis = slender
multifidus = much divided
obliquus = lopsided, slanting, oblique
ortho- (prefix) = upright, straight
patens = spreading
pendulus = hanging, weeping
procumbens = prostrate
procurrens = spreading, running
profusus = abundant
prostratus = prostrate
pyramidalis = pyramid-shaped
ramosus = branched
ramulosus = twiggy
rectus = upright
repens, reptans = creeping
rigens, rigidus = stiff, rigid
rigescens, rigidulus = rather stiff or rigid

scandens = climbing
simplex = unbranched
soboliferus = with creeping stems that form roots
socialis = growing in colonies
spiralis = spiral, corkscrewed
stans = upright
stragulus, stragulatus = mat-forming
strictus = upright
suffruticosus = somewhat shrubby
suspensus = hanging, weeping
tenuis, tenuissimus = thin, slender, fine
tortuosus, tortus, tortilis = contorted
tristis = weeping
vescus = small, feeble
virgatus = twiggy
viticella = a small vine
volubilis = winding, twining

SIZE

altissimus = very tall
altus = tall
ampliatus = enlarged
amplissimus = very large
amplus = large
aperantus = of unlimited growth
decumanus = very large
diminutus = small
elatior = taller
elatus = tall
excelsus, excelsior = tall, taller

giganteus = very large, giant

gigas = giant

grandi- (prefix) = large

grandis = large

humilis = low-growing, dwarf

imperialis = large, showy, stately

macr- (prefix) = large, or (occasionally) long

magni- (prefix) = large

magnus = large

major = greater, larger

maximus = largest

medius = medium

mega-, megalo- (prefixes) = very large

micr- (prefix) = small

minimus = smallest

minor = smaller, lesser

minutissimus = very small indeed

minutus = very small

nanus = dwarf

nobilis = large or noble

parvi- (prefix) = small

parvus = small

procerus = very tall

pumilus, pumilio = dwarf

pygmaeus = very small

LEAF SHAPE AND ARRANGEMENT

acerosus, acicularis = needle-like

acinaceus = scimitar-shaped

acu- (prefix) = pointed

acuminatus = tapering to a point

aduncus = hooked

aesculifolius = foliage like a horse chestnut

alatus = winged

alternifolius = with alternate leaves

anethifolius = with dill-like leaves

angulidens = having hooked teeth

anguliger = having hooks

angustatus = narrowed

angustifolius = narrow-leaved

apifolius = with celery-like leaves

aquifolius = with holly-like leaves

argutus = sharply toothed or notched

arundinaceus = reed-like

attenuatus = narrowing to a point

belophyllus = with arrow-
   shaped leaves

blepharophyllus = leaves
   fringed like eyelashes

brachy- (prefix) = short

brevis and *brevi-* (prefix) =
   short

calamifolius = with reed-like
   leaves

cannabinus = with hemp-like
   leaves

capillaceus = slender, hair-like

clavatus = club-shaped

cordatus = heart-shaped

cordifolius = with heart-shaped
   leaves

crispus = curled

cuneatus, cuneifolius,
   cuneiformis = wedge-like

didymus = paired, two-lobed

ensatus, ensifolius, ensiformis =
   sword-like

falcatus, falciformis,
   falcifolius, falcinellus =
   sickle-like

flabellatus, flabelliformis =
   fan-shaped

hastatus = spear-shaped

integrifolia = with undivided
   leaves

lanceus, lanceolatus =
   spear-like

latifolius = broad-leaved

ligularis, ligulatus = strap-like

lingua, lingulatus, linguiformis
   = tongue-shaped

linifolius = with flax-like leaves

longifolius = long-leaved

lorifolius = with strap-like
   leaves

lyratus = lyre-shaped

macrophyllus = large-leaved

microphyllus = small-leaved

myrtifolius = with myrtle-like
   leaves

orbicularis = round and flat

ovatus = egg-shaped

palmatus = hand-shaped

panduratus = fiddle-shaped

peltatus = shield-like

perfoliatus = with leaves
   surrounding the stem

petiolaris = with long-stalked
   leaves

pinnatus = with pinnate leaves

platyphyllus, platyphyllos =
   broad-leaved

pungens = sharply pointed

rotundifolius = round-leaved

sagittalis, sagittifolius =
   arrow-shaped

salicifolius = with willow-like
   leaves

sessilis = without a stalk

sonchifolia = with thistle-like leaves

spathulatus = spatula-shaped

stenophyllus = narrow-leaved

subulatus = awl-shaped

tenuifolius = fine-leaved

trifolius, trifoliatus = three-leaved

verticillatus = with whorled leaves

vitifolius = with grapevine-like leaves

LEAF MARGINS

biserratus = double-toothed

crenatus = round-toothed

dentatus = toothed

denticulatus = fine-toothed

fimbriatus = fringed

laciniatus = deeply cut

repandus = slightly wavy

serratus, serratifolius = saw-toothed, serrated

serrulatus = with smaller teeth

undulatus = wavy

LEAF TEXTURE

bullatus = blistered or puckered

callosus = thick skinned

calophyllus = beautiful-leaved

calvus = hairless

caperatus = wrinkled

capillatus = finely haired

costatus = strongly ribbed

crassifolius = thick-leaved

glabrus = smooth, hairless

glutinosus = sticky

laevigatus = smooth

laevis = soft

lanatus = woolly

mollis = soft, velvety

nervosus = prominently veined

nitens, nitidus = glossy, shining

pubescens = downy

rugosus = wrinkled

tomentosus = downy

villosus = softly hairy

LEAF MARKINGS

areolatus = marked in small circular areas

marginalis, marginatus = with coloured margins

pictus = blotched, mottled

zonalis = with banding

APPLIED TO FLOWERS OR FOLIAGE

ciliaris, ciliatus, cilosus = fringed like eyelashes

guttatus = speckled

maculatus = spotted

punctatus = spotted, dotted

reticulatus = netted

sericeus = silken

striatulus = rather striped

striatus = striped

FLOWERS

acaulis = stemless

angustiflorus = narrow-flowered

angustipetalus = narrow-petalled

anopetalus = with erect petals

barbatus = bearded

barbiger, barbigerus = bearded, hairy

calcaratus = spurred

callianthus = having beautiful flowers

campanulatus, campaniflorus = with bell-shaped flowers

capitatus = with a dense head of flowers

capitellatus = small-headed

caulescens = with a stem

cernuus = nodding, drooping

clausus = closed

coronarius = garland-like, crown-like

densiflorus = densely flowered

dolicanthus = long-flowered

flore-pleno = double-flowered

floribundus = profusely flowering

floridus = free flowering

globosus = spherical, round

glomeratus = cluster-headed

grandiflorus = large-flowered

involucratus = having a circle of bracts around flowers

laxiflorus = loose-flowered

longiflorus = with long-shaped flowers

longipes = long-stalked

macranthus = large-flowered

magniflorus = large-flowered

micranthus = small-flowered

non-scriptus = unmarked

nudiflorus = with flowers appearing before the leaves

nutans = nodding

paniculatus = panicled

parviflorus = small-flowered

pauciflorus = with few flowers

plenus = double-flowered

plumosus = feathery

racemosus = with flowers in racemes

reflexus = with petals bent back

sphaerocephalus = round-headed

spicatus = spiked

stellatus = star-like

torquatus = collared

umbellatus = with flowers in umbels

BULBS, STALKS, SHOOTS AND ROOTS

articulatus = jointed

bulbiferus = bulb-bearing, producing bulbils

cirrhosus = with tendrils

deltoides = triangular

macrorhizus = large-rooted

napellus = tubers like small turnips

psilostemon = smooth-stemmed

stoloniferus = with root runners

triquetrus = three-sided

tuberosus = with tuberous roots

viminalis, vimineus = with long, thin shoots

PLANT USES

coronarius = used for making wreaths and garlands

domesticus = domesticated

edulus, esculentus = edible

frumentaceus = grain bearing

funebris = used for funerals or in cemeteries

oleraceus, holeraceus = relating to vegetables or kitchen garden

papyriferus = paper-bearing

ritualis = used in rituals

sativus = cultivated

viniferus = wine producing

GENERAL

admirabilis = worthy of notice

amabilis = lovely

amoenus = lovely

annuus = annual, to distinguish from biennial or perennial species

augustus = venerable, majestic

basilicus = princely, royal

benedictus = blessed, of good reputation

biennis = biennial, to distinguish from annual or perennial species

blandus = pleasing, charming; mild

callimorphus = with beautiful shape

callistachyus = with beautiful spikes

callistus = very beautiful

callizonus = beautifully banded

communis = common, general, growing in company

confusus = uncertain, easily mistaken for another plant

decorus, decoratus = decorative

dulcis = sweet

elegans = elegant

elegantissimus = very elegant

erromenus = vigorous, healthy

eucharis = pleasing, agreeable

eudoxus = of good reputation

eximius = distinguished, extraordinary

facetus = fine, choice

formosissimus = very beautiful

formosus = beautiful

fortis = strong

generosus = noble, eminent

gloriosus = glorious

gracilis = graceful

gracillimus = most graceful

insignis = handsome, distinguished

jucundus = pleasant, delightful

magnificus = splendid, magnificent

mirabilis = marvellous

monstrosus = monstrous

nobilis = noble, but also large

ornatus = ornate

perennis = perennial, to distinguish from annual or biennial species

praestans = distinguished, excellent

princeps = princely

pulchellus = beautiful but small

pulcher = beautiful

pulcherrium = very beautiful

regalis = regal, stately

robustus = robust, strong

speciosus = showy, handsome

spectabilis = spectacular, showy

splendens = splendid

splendidum = very splendid

superbus = superb

tenax = tough

validus = strong, robust

venustus = lovely, pleasing

vulgaris = common

MISCELLANEOUS

affinis = related to

dioicus = with male and female organs on separate plants

heterophyllus = with different kinds of leaves on the same plant

hybridus = hybrid

intermedius = hybrid that is intermediate between parent plants

sempervirens = evergreen

umbrosus = shade-loving

BIBLIOGRAPHY

My interest in the naming of plants was initiated by two little books, both of them written a long time ago and picked up very inexpensively at the RHS's Malvern Spring Gardening Show (as it was then called). G. F. Zimmer's *A Popular Dictionary of Botanical Names and Terms with Their English Equivalents* was published by Routledge & Kegan Paul in 1912, and A. T. Johnson's *Plant Names Simplified: Their Pronunciation, Derivation & Meaning* was originally published by W. H. & L. Collingridge in 1931, then revised and expanded by Henry A. Smith in 1946. Although many of the names listed have since changed, these books have remained an inspiration. The other two essential volumes that were never off my desk were those by W. T. Stearn itemised below, but all the books in the following list have been immensely helpful and are recommended to anyone with an interest in plants, their naming and their history.

Arber, Agnes, *Herbals: Their Origin and Evolution*,
      Cambridge University Press, 1912 (rev. edn, 1938)
Blunt, Wilfrid, *The Compleat Naturalist: A Life of Linnaeus*,
      Collins, 1971
Bowles, E. A., *My Garden in Spring*, T. C. & E. C. Jack, 1914
——, *My Garden in Summer*, T. C. & E. C. Jack, 1914
——, *My Garden in Autumn and Winter*, T. C. & E. C.
      Jack, 1915

Brickell, Christopher, *The RHS Gardener's Encyclopaedia of Plants and Flowers*, Dorling Kindersley, 1989

Brittain, Julia, *Plant Names Explained*, David & Charles, 2005

——, *Plants, People & Places*, David & Charles, 2006

Brooke, Jocelyn, *The Wild Orchids of Britain*, The Bodley Head, 1950

——, *The Flowers in Season*, The Bodley Head, 1952

Brown, Jane, *The Pursuit of Paradise*, HarperCollins, 1999

Bruton-Seal, Julie, and Seal, Matthew, *The Herbalist's Bible: John Parkinson's Lost Classic Rediscovered*, Merlin Unwin, 2014

Campbell-Culver, Maggie, *The Origin of Plants*, Headline, 2001

Coats, Alice M., *Flowers and their Histories*, Hulton Press, 1956

——, *Garden Shrubs and their Histories*, Vista Books, 1963

——, *The Quest for Plants*, Studio Vista, 1969

——, *The Book of Flowers*, Chancellor Press, 1973

——, *The Treasury of Flowers*, Phaidon, 1975

Cowell, John, *The Curious and Profitable Gardener*, 1730

Delbourgo, James, *Collecting the World: The Life and Curiosity of Hans Sloane*, Allen Lane, 2017

Farrer, Reginald, *My Rock-Garden*, Edward Arnold, 1908

——, *The English Rock-Garden* (2 vols), Edward Arnold, 1919

Finlay, Victoria, *Colour*, Hodder & Stoughton, 2002

Gerard, John, *The Herball or Generall Historie of Plantes* (revised by Thomas Johnson, 1633), Dover, 1975

Gledhill, David, *The Names of Plants*, 4th edn, Cambridge University Press, 2008

Gordon, James, *The Planters, Florists, and Gardeners Pocket Dictionary*, 1774

Gribbin, Mary and John, *Flower Hunters,* Oxford University Press, 2008

Griffiths, Mark, *RHS Index of Garden Plants,* Macmillan, 1994

Grigson, Geoffrey, *The Englishman's Flora,* Phoenix House, 1955

——, *A Dictionary of English Plant Names,* Allen Lane, 1974

Gunther, R. T., *Early British Botanists and Their Gardens,* Oxford University Press, 1922

Hamilton, Jill, Duchess of, and Bruce, Julia, *The Flower Chain,* Kangaroo Press, 1998

Harrison, Lorraine, *RHS Latin for Gardeners,* Mitchell Beazley, 2012

Harvey, John, *Early Gardening Catalogues,* Phillimore, 1972

——, *Early Nurserymen,* Phillimore, 1974

Henrey, Blanche, *British Botanical and Horticultural Literature before 1800* (3 vols), Oxford University Press, 1975

Jackson, B. Daydon, *A Glossary of Botanic Terms,* Gerald Duckworth, 1900

Johns, C. A., *Flowers of the Field,* SPCK, (1853) 29th edn revised by G. S. Boulger, 1899

Johnson, Thomas (ed. J. S. L. Gilmour), *Botanical Journeys in Kent & Hampstead,* The Hunt Botanical Library, 1972

Linnaeus, Carl, *Species Plantarum* (1753), 2 vols, The Ray Society, 1959

Mabey, Richard, *Flora Britannica,* Sinclair-Stevenson, 1996

McClintock, David, *Companion to Flowers,* G. Bell & Sons, 1966

Mayhew, Henry, *London Labour and the London Poor* (1861) 4 vols, Dover, 1968

Miller, Philip, *The Gardeners Dictionary*, 8th edn, 1768

Morwood, James (ed.), *Pocket Oxford Latin Dictionary*, Oxford University Press, 2005

Oakley, Henry, *Doctors in the Medicinal Garden*, Royal College of Physicians, 2012

Page, Martin (foreword), *Name That Plant*, Worth Press, 2001

Parkinson, John, *Paradisi in Sole Paradisus Terrestris* (1629), Dover, 1976

——, *Theatrum Botanicum*, 1640

Pavord, Anna, *The Naming of Names*, Bloomsbury, 2005

Pliny the Elder (trans. and ed. John F. Healey), *Natural History: A Selection*, Penguin, 1991

Pratt, Anne, *Wild Flowers* (2 vols), SPCK, 1865–6

Raven, Charles E., *John Ray, Naturalist*, Cambridge University Press, 1942

——, *English Naturalists from Neckham to Ray*, Cambridge University Press, 1947

Raven, John E., *A Botanist's Garden*, Silent Books, 1971

——, *Plants and Plant Lore in Ancient Greece*, Leopard's Head Press, 2000

Ray, John (trans. A. H. Ewen and T. C. Prime), *Ray's Flora of Cambridge* (1660), Wheldon and Wesley, 1975

Rothschild, Miriam, *The Rothschild Gardens*, Gaia, 1996

St Clair, Kassia, *The Secret Life of Colour*, John Murray, 2016

Sanders, T. W., *Sanders' Encyclopaedia of Gardening* (1895), rev. edn, W. H. & L. Collingridge, 1931

Shulman, Nicola, *A Rage for Rock Gardening*, Short Books, 2001

Stacey, Robyn, and Hay, Ashley, *Herbarium*, Cambridge University Press, 2004

Stearn, William T., *Botanical Latin: History. Grammar, Syntax, Terminology and Vocabulary*, David & Charles, (1966) rev. edn, 1983

——, *Stearn's Dictionary of Plant Names for Gardeners*, Cassell, (1972) rev. edn, 2004

Stuart, David, *Dangerous Garden*, Frances Lincoln, 2004

Stuart, David, and Sutherland, James, *Plants from the Past*, Viking, 1987

Theophrastus (trans. A. F. Hort), *Enquiry into Plants* (2 vols), Heinemann & Harvard University Press, 1961

Thomas, Graham Stuart, *Perennial Garden Plants*, J. M. Dent, 1976

——, *Ornamental Shrubs, Climbers & Bamboos*, John Murray, 1992

Tournefort, Joseph Pitton de (trans. John Ozell), *A Voyage into the Levant* (2 vols, 1718), Cambridge University Press, 2014

Turner, William, *A New Herball* (1551), The Mid Northumberland Arts Group and Carcanet Press, 1989

——, *Libellus de Re Herbaria 1538 / The Names of Herbes 1548: Facsimiles*, The Ray Society, 1965

Veitch, James H., *Hortus Veitchii*, James Veitch & Sons, 1906

White, T. H., *The Book of Beasts*, Jonathan Cape, 1954

A number of the earliest books are available online at such invaluable websites as the Internet Archive (www.archive. org) and the Biodiversity Heritage Library (www.biodiversitylibrary.org).

Correct botanical plant names can be checked on the RHS Horticultural Database (http://apps.rhs.org.uk/ horticulturaldatabase).

Other online resources I have found particularly useful:

W.J. Bean's *Trees and Shrubs Hardy in the British Isles*:
    www.beanstreesandshrubs.org
Global Plants: https://plants.jstor.org
The Oxford Dictionary of National Biography:
    www.oxforddnb.com

ACKNOWLEDGEMENTS

My first thanks go to David Wheeler, who many years ago gave me the opportunity to write about plant names (and a good deal else) in his gardening quarterly *HORTUS*. I'd also like to thank those other editors who over time commissioned me to write about plants and gardens, notably Tiffany Daneff and Kylie O'Brien at the *Daily Telegraph*.

My former agent, the late and much missed David Miller, did much to make this book happen, and I am also very grateful to his successor, Natasha Fairweather, who has been a staunch support while the book actually got written and went through to press. At Little, Brown, Richard Beswick once again proved the best of editors, and my thanks also go to Tamsyn Berryman, Marie Hrynczak, Charlotte Stroomer, Zoe Hood and Kim Nyamhondera. If not precisely Thomas Johnson to my John Gerard, Elizabeth Dobson nevertheless used her sharp eye and her considerable knowledge to deal with an inexcusably messy typescript, and she has saved me from innumerable typing errors, incorrect renderings of Latin and Greek words, and other egregious blunders. My debt to her is immeasurable, and any mistakes that remain are my own. Without the astonishing resources and knowledgeable staff of the London Library this book would have been impossible to write.

For all manner of things, including books and bits of information but not always so easily quantifiable, I am very grateful to Adam Bager, Prasun Banerjee, Edward Behrens, Tamsyn Blaikie, Andreas Campomar, Niladri Chatterjee, Naman Chaudhary, Simon Dorrell, Georgina Hammick, Lance and Jane Hattatt, John Matheson, Candia McWilliam, Brian Oatley, Sue Nevill-Parker, the late Pat Parker, Alice Sielle and Salley Vickers. As always, my greatest thanks go to Christopher Potter, with whom I have made two London gardens, and whose enthusiasm when reading the book in typescript was just what I needed at a crucial stage in the writing of it.

By his example and practice, Thomas Blaikie transformed what had been a general but uninformed interest in plants into something much more focused and passionate. In over forty years of friendship he has been a knowledgeable and enlivening companion on innumerable trips to gardens and nurseries. This book is for him.

Peter Parker, London E3, June 2018

ILLUSTRATION CREDITS

Illustrations reprinted from *200 Illustrations from Gerard's Herbal* (Dover Publications, 2005) on pp. 3, 36, 60, 64, 84, 143, 202.

Illustration reprinted from *The Herbal or General History of Plants* by John Gerard (Dover Publications, 1975) on p.4.

Illustrations reprinted from *Gerard's Herball: The Essence thereof distilled by Marcus Woodward from the Edition of Th. Johnson, 1636*, by John Gerard and printed by R. & R. Clark, Ltd. Edinburgh for Gerald Howe (1927) on pp. 11, 16, 18, 28, 58, 61, 69, 106, 108, 120, 131, 135, 136, 138, 142, 145, 150, 157, 164, 173, 176, 180, 185, 189, 192, 194, 209, 220, 225, 227, 229, 231, 269, 273, 275, 278, 283, 289, 291, 297, 302.